Mathematical models for the growth of human populations

Mathematical models for the growth of human populations

J.H.POLLARD
MACQUARIE UNIVERSITY, SYDNEY, AUSTRALIA

CAMBRIDGE UNIVERSITY PRESS
CAMBRIDGE
LONDON · NEW YORK · MELBOURNE

CAMBRIDGE UNIVERSITY PRESS
Cambridge, New York, Melbourne, Madrid, Cape Town,
Singapore, São Paulo, Delhi, Mexico City

Cambridge University Press
The Edinburgh Building, Cambridge CB2 8RU, UK

Published in the United States of America by Cambridge University Press, New York

www.cambridge.org
Information on this title: www.cambridge.org/9780521294423

First published 1973
Reprinted 1975
Re-issued 2013

A catalogue record for this publication is available from the British Library

ISBN 978-0-521-20111-7 Hardback
ISBN 978-0-521-29442-3 Paperback

To my mother and my father

Contents

Preface

This book is an account of some of the mathematical models which have been proposed for studying the growth of human populations, and the emphasis is on stochastic models, i.e. systems that change in accordance with probabilistic laws. A few of these models are mentioned in other texts, but many of the more complex models appear only in original papers in diverse journals. The book began in the form of lecture notes prepared for graduate students in the Department of Statistics at the University of Chicago in the Spring Quarter of 1968, and experience gained at the University of Chicago and later at Macquarie University has suggested numerous modifications and additions to the original text.

The aim has been to make the book as self-contained as possible, and therefore the mathematical techniques used have been kept as elementary as possible. They are mostly standard undergraduate results of calculus, matrix algebra and basic statistics. There are a few instances in which rather more complicated methods are necessary, or less well-known techniques are employed, and then detailed explanations are given. One very important aspect of the subject has not been treated (except marginally): the estimation of numerical values for the life table. Specialist actuarial techniques are usually employed, and these numerical methods are described elsewhere.

The book is intended primarily as a reference work for demographic research workers, but it might be used as a textbook for a final-year undergraduate mathematics class or graduate demography students with strong backgrounds in mathematics.

Certain sections have been marked with asterisks: the material contained in them is not required for an understanding of later chapters, and they may be safely omitted, at least at a first reading. Exercises are given at the end of each chapter, and in solving these problems the student should learn a great deal more about the basic mathematical models involved. Solutions for these exercises are outlined on pages 158–174.

Fragments of the history of the subject are given in the main body of the text, and certain interesting historical references are quoted. References are given on pages 150–154.

I wish to thank Nathan Keyfitz for inviting me to the University of Chicago in 1967 and for the inspiration he has given me. I am indebted to Dr C. C. Heyde and Dr E. Seneta for telling me about Bienaymé's work in 1845 on the simple branching process and for allowing me to see the manuscript of their paper. Dr C. D. Cooper very kindly read an earlier version of chapter 4 and helped eliminate a major error in section 4.3; he also suggested the matrix in the final exercise of that chapter. Thanks are due also to Miss Jane Blaxland and Miss Helen Knight for typing several versions of the manuscript, and to Miss Betty Thorne for drawing the diagrams.

Finally, I should like to thank my wife for her patience during the many evenings when I was busy with this book.

Macquarie University, Sydney J.H.P.
January 1973

NOTE to 1975 IMPRESSION: Certain errors have come to light since this book was first published. These have now been corrected, and I wish to thank Graham Pollard for bringing some of these errors to my notice. I am also indebted to Professor H. O. Lancaster for drawing my attention to the work of Leonhard Euler (1707–83).

J.H.P.

1
Introduction

The leaders and governments of most civilizations have collected statistical information about their peoples. The Sumerians, five thousand years ago, for example, enumerated their population for taxation purposes. Later, the Romans based conscription on enumeration of the population. The Christian Church has over many centuries compiled, in the form of parish registers, a huge amount of demographic material.[1] Today, the quantity of data available about current populations, even in some of the less-developed countries, is enormous.

It was only in the seventeenth century, however, that men became interested in studying the numbers of their fellow human-beings from the purely scientific point of view. The Englishman, John Graunt (1620–74) was probably the first of these, and his work was thorough and of a high standard (J. Graunt, 1662[2]; I. Sutherland, 1963). He produced the first life table and studied the population of London in some detail. Many others were to follow his example.

We are primarily concerned with mathematical models for human populations. Leonhard Euler produced such a model as early as 1767[3],and the population analysis of Thomas Malthus (1798[2]) could

[1] Several groups are currently studying the data available from parish registers. Those at the Institute of Genetics, Pavia, Italy are interested in genetic data obtainable from parish records in the Parma Valley, Italy (L. L. Cavalli-Sforza, 1958). The Cambridge Group for the History of Population and Social Structure are interested mainly in demographic results from pre-industrial populations in England (E. A. Wrigley, 1966; E. A. Wrigley and R. S. Schofield, 1968). Some of the better-known work has been done in France (P. Goubert, 1960).

[2] These references are extremely interesting (and sometimes amusing) reading.

[3] 'Recherches générales sur la mortalité et la multiplication du genre humain'. *Histoire de l'academie royale des sciences et belles-lettres, Année 1760*, pp. 144–64.' Preussische Akademie der Wissenschaften zu Berlin. The date of publication is 1767. See also I. Todhunter (1865). *A History of the Mathematical Theory of Probability*, pp. 240–1. (Reprinted in 1965 by Chelsea Publishing Company, New York.)

be described as the formulation of a mathematical model. Modern population theory, however, might be said to have begun with the deterministic theories of A. J. Lotka (1907) and F. R. Sharpe and A. J. Lotka (1911).

An account of the life table and its use is given in chapter 2. The next eight chapters are devoted to various mathematical models which have been proposed for analysing the growth of human populations. These models vary considerably in their complexity and in their mathematical treatment.

2

The life table

2.1 *Introduction*

Consider a large number n_0 of lives all aged exactly 0 years. After x years, some of these lives will have died, and there will be, say, n_x lives remaining, all aged x. It is clear that n_x forms monotonic non-increasing sequence of integers with increasing x. If n_x is large, the probability of survival from age x to age $x+t$, denoted by ${}_tp_x$, will be approximately equal to n_{x+t}/n_x. That is,

$$ {}_tp_x \doteqdot n_{x+t}/n_x. \tag{2.1.1}$$

This is the type of argument first used by demographers to develop the important theoretical concept of the *life table*. The first life table, compiled by John Graunt in 1662, was mentioned in chapter 1. This is commonly regarded as his masterpiece. The following quotation comes from his book:

> '9. Whereas we have found, that of 100 quick Conceptions about 36 of them die before they be six years old, and that perhaps but one surviveth 76, we, having seven Decads between six and 76, we sought six mean proportional numbers between 64, the remainder, living at six years, and the one, which survives 76, and finde, that the numbers following are practically near enough to the truth; for men do not die in exact Proportions, nor in Fractions: from whence arises this Table following.

Viz. of 100 there dies within the first six years	36
The next ten years, or Decad	24
The second Decad	15
The third Decad	09
The fourth	6
The next	4
The next	3
The next	2
The next	1

'10. From whence it follows, that of the said 100 conceived there remains alive at six years end 64.

At Sixteen years end	40
At Twenty six	25
At Tirty six	16
At Fourty six	10
At Fifty six	6
At Sixty six	3
At Seventy six	1
At Eighty	0'

In spite of its humble origin, the modern life table is a fine theoretical tool. A continuous, well-behaved, monotonic decreasing function l_x is defined such that the probability of survival from exact age x to exact age $x+t$, ${}_tp_x$ is equal to l_{x+t}/l_x. Some arbitrary value like 10,000 or 100,000 is usually assigned to l_0, and this value is called the *radix*.

Although l_x is continuous and is defined for all values of x greater than zero, it is an empirical function and seldom (if ever) has an explicit mathematical form. It is usually tabulated for integral values of x, and intermediary values (if required) are found by interpolation. The difference $l_x - l_{x+1}$ represents the number of deaths aged x last birthday out of l_x lives who attain exact age x, and it is denoted by d_x.

The life table for Australian males (1961) is given on pages 175–177.

2.2 *Mortality computations*

Using the life table, it is possible to compute various probabilities involving mortality. Consider, for example, a life now aged 30. What is the probability that this life will die between exact age 40 and exact age 50?

According to our definition of l_x, the probability that a life aged 30 will survive to exact age 40 is l_{40}/l_{30}, and the probability that a life aged 40 will die before exact age 50 is $1-(l_{50}/l_{40})$. The mortality prospects of an individual at different ages are assumed independent, and we conclude that the life aged 30 will die between ages 40 and 50 with probability $(l_{40}/l_{30})(1-l_{50}/l_{40})$. This probability simplifies to $(l_{40}-l_{50})/l_{30}$, and for the Australian life table (males) 1961 it is equal to $(92{,}859-88{,}473)/94{,}726 = 0.046$.

2.3 *The force of mortality*

An example of the use of the life table function l_x for mortality computations is given in section 2.2. Consider now a life aged x. What is the probability that this life will die between exact age $x+t$ and exact age $x+t+\mathrm{d}t$? An argument similar to that given in section 2.2 indicates that the probability is $(l_{x+t}-l_{x+t+\mathrm{d}t})/l_x$. The function l_x is well-behaved and $l_{x+t+\mathrm{d}t}$ may be expanded in a Taylor series about the point $x+t$. It follows that

$$\frac{l_{x+t}-l_{x+t+\mathrm{d}t}}{l_x} = \frac{l_{x+t}}{l_x}\left\{-\frac{1}{l_{x+t}}\frac{\mathrm{d}}{\mathrm{d}t}(l_{x+t})\,\mathrm{d}t\right\}+o(\mathrm{d}t)$$

$$= {}_tp_x\mu_{x+t}\mathrm{d}t+o(\mathrm{d}t), \tag{2.3.1}$$

where

$$\mu_x = -\frac{1}{l_x}\frac{\mathrm{d}}{\mathrm{d}x}l_x = -\frac{\mathrm{d}}{\mathrm{d}x}\log l_x. \tag{2.3.2}$$

Formula (2.3.1) should be noted: ${}_tp_x$ is the probability that a life survives from age x to age $x+t$, and $\mu_{x+t}\mathrm{d}t$ is the probability that a life aged $x+t$ will die during the time element $\mathrm{d}t$. The mortality function μ_x, defined by equation (2.3.2), is usually referred to as the *force of mortality at age x*. It is of considerable theoretical importance.

2.4 *Numerical evaluation of the force of mortality*

Numerical values of μ_x are often required. The function l_x is usually an empirical function, so numerical differentiation is necessary to determine μ_x. The following formula is frequently employed by the compilers of life tables, and it is accurate provided l_x is a polynomial of fourth degree in the vicinity of x:

$$\mu_x = \frac{8(l_{x-1}-l_{x+1})-(l_{x-2}-l_{x+2})}{12l_x}. \tag{2.4.1}$$

It may be derived by expanding l_{x-2}, l_{x-1}, l_{x+1} and l_{x+2} in Taylor series about the point x and eliminating the terms involving second, third and fourth derivatives of l_x.

It is not possible to compute μ_0 and μ_1 using this formula. The force of mortality μ_x is changing very rapidly in the age range 0 to 2, so any formula using values of l_x for integral x only is unlikely to give reliable results. The problem of computing μ_x for $0 \leqslant x \leqslant 2$ is extremely complex, and in the Australian life table (males) for 1961, for example, μ_x is only given for ages greater than two years. Approximate values can be obtained by fitting a hyperbola to the l_x function in this age range.

2.5 *Other life table functions*

The life table functions l_x, μ_x, and ${}_tp_x$ are defined above. Other life table functions are defined in terms of l_x, and we shall have occasion to use some of them in later chapters:

$$d_x = l_x - l_{x+1} \quad \text{(deaths)},$$

$$_np_x = l_{x+n}/l_x,$$

$$p_x = {}_1p_x,$$

$$_nq_x = 1 - {}_np_x,$$

$$q_x = 1 - p_x \quad \text{(mortality rate)},$$

$$L_x = \int_0^1 l_{x+t}\,dt,$$

$$m_x = d_x/L_x \quad \text{(central mortality rate)}.$$

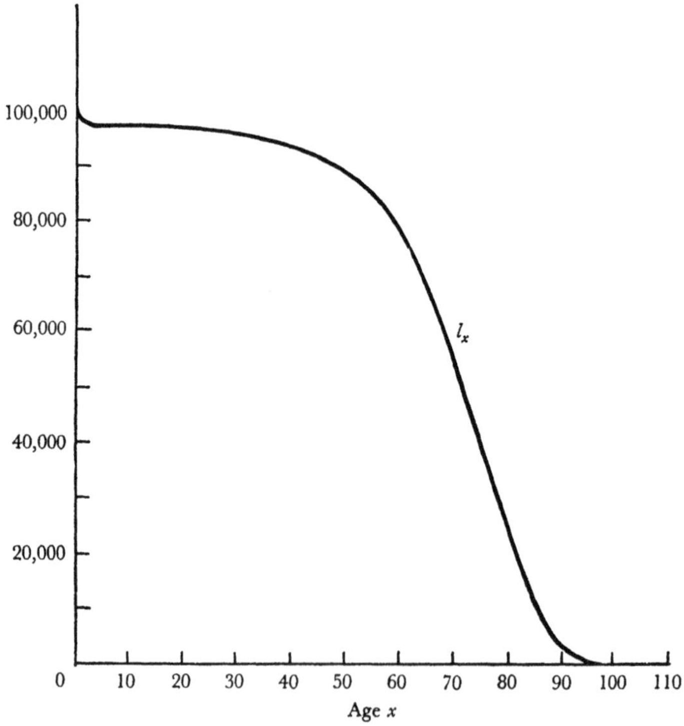

Figure 2.6.1. Graph of l_x (Australian life table (males) 1961).

2.6 *The graphs of certain life table functions*

The graphs of l_x, μ_x, and q_x for Australian Males (1961) are given in figures 2.6.1, 2.6.2 and 2.6.3 respectively. The following comments about μ_x help in the understanding of these curves:

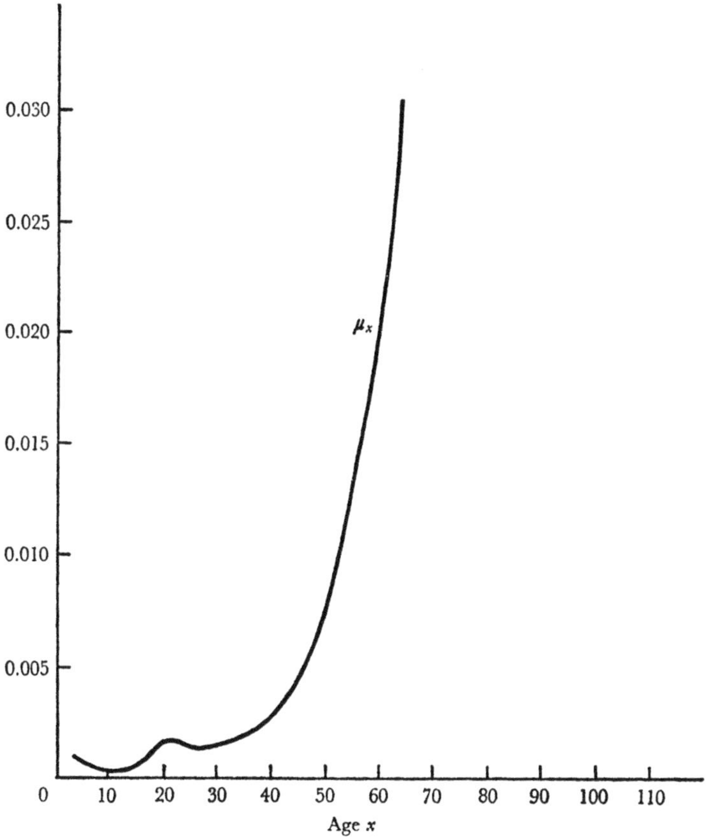

Figure 2.6.2. Graph of μ_x (Australian life table (males) 1961).

(i) μ_0 is large.
(ii) It drops rapidly in the first two years of life.
(iii) It is a minimum near 11 years of age.
(iv) It rises *very* gradually to reach a maximum in the early twenties.
(v) It then drops a little before continuing its gradual ascent.

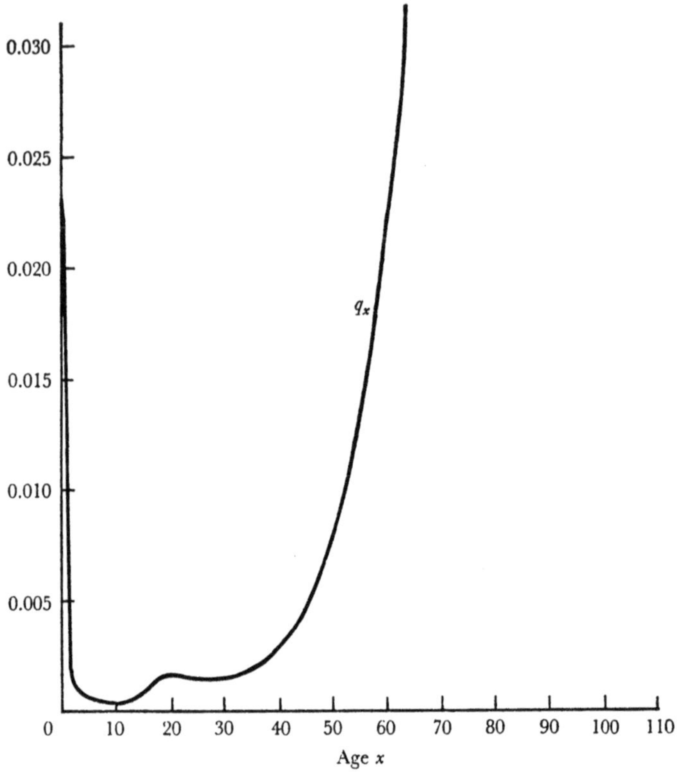

Figure 2.6.3. Graph of q_x (Australian life table (males) 1961).

The maximum value in the early twenties is caused by motor accidents, etc.

(vi) After age 50, μ_x starts to rise more rapidly.

The q_x curve is of course very similar to the μ_x curve.

2.7 *The Euler–Maclaurin expansion*

The Euler–Maclaurin expansion connects the integral of a function with the sum of a series of equally spaced ordinates. To prove the formula, we make use of the differential operator D, and the finite difference operators Δ and E. These three operators are defined as follows:

$$\left.\begin{aligned} \mathrm{D}U_x &= \frac{\mathrm{d}}{\mathrm{d}x}U_x, \\ \Delta U_x &= U_{x+1} - U_x, \\ \mathrm{E}U_x &= U_{x+1}. \end{aligned}\right\} \tag{2.7.1}$$

Relationships exist between the operators, and the following two are important in the present context:

$$\left.\begin{aligned} E &\equiv 1+\Delta; \\ E &\equiv e^{D}. \end{aligned}\right\} \qquad (2.7.2)$$

The first relationship is very easy to prove, and the second may be derived by considering the Taylor series expansion:

$$\begin{aligned} EU_x &= U_{x+1} \\ &= U_x + U_x' + \frac{1}{2!}U_x'' + \frac{1}{3!}U_x''' + \dots \\ &= \left(1 + D + \frac{1}{2!}D^2 + \frac{1}{3!}D^3 + \dots\right)U_x \\ &= e^{D}U_x. \end{aligned}$$

To derive the Euler–Maclaurin expansion, we consider the sum

$$\begin{aligned} U_0 + U_1 + U_2 + \dots + U_{n-1} &= (E^0 + E^1 + E^2 + \dots + E^{n-1})\,U_0 \\ &= (E^n - 1)(E-1)^{-1}U_0 \\ &= (E^n - 1)(e^{D}-1)^{-1}U_0 \\ &= (E^n - 1)\left(\frac{1}{D} - \frac{1}{2} + \frac{1}{12}D + \dots\right)U_0. \end{aligned}$$

That is $$\sum_{0}^{n-1} U_i = \int_0^n U_x\,dx - \tfrac{1}{2}(U_n - U_0) + \tfrac{1}{12}(U_n' - U_0') + \dots \qquad (2.7.3)$$

and this is the Euler–Maclaurin expansion.

2.8 *The expectation of life*

The *curtate expectation of life* is the average number of *whole* years lived after age x by a life who attains age x. It is denoted by e_x. Clearly,

$$\begin{aligned} e_x &= \sum_{n=0}^{\infty} n\frac{d_{x+n}}{l_x} \\ &= \frac{1}{l_x}\{(l_{x+1}-l_{x+2}) + 2(l_{x+2}-l_{x+3}) + 3(l_{x+3}-l_{x+4}) + \dots\} \\ &= \frac{1}{l_x}\sum_{n=1}^{\infty} l_{x+n} \\ &= \frac{1}{l_x}\sum_{n=0}^{\infty} l_{x+n} - 1. \end{aligned} \qquad (2.8.1)$$

The *complete expectation of life* is the average number of years of life lived after age x by a life who attains age x. It is denoted by $\mathring{e}_x$. The probability that a life aged x will die at age $x+t$ is given by equation (2.3.1), and it follows that the expectation of life at age x is given by

$$\begin{aligned}\mathring{e}_x &= \frac{1}{l_x}\int_0^\infty t l_{x+t}\mu_{x+t}\,\mathrm{d}t \\ &= -\frac{1}{l_x}\int_0^\infty t\frac{\mathrm{d}}{\mathrm{d}t}(l_{x+t})\,\mathrm{d}t \\ &= \frac{1}{l_x}\int_0^\infty l_{x+t}\,\mathrm{d}t \quad \text{(integration by parts).}\end{aligned} \tag{2.8.2}$$

By general reasoning, it seems plausible that $\mathring{e}_x$ should be greater than e_x by about half a year. The Euler–Maclaurin expansion does in fact provide a relationship between the two expectations. When applied to formula (2.8.1),

$$\begin{aligned}e_x + 1 &= \frac{1}{l_x}\int_0^\infty l_{x+t}\,\mathrm{d}t + \frac{1}{l_x}(\tfrac{1}{2}l_x) - \frac{1}{12}\frac{1}{l_x}(l_x') + \ldots \\ &= \mathring{e}_x + \tfrac{1}{2} + \tfrac{1}{12}\mu_x + \ldots .\end{aligned}$$

Hence

$$\mathring{e}_x \doteqdot e_x + \tfrac{1}{2} - \tfrac{1}{12}\mu_x. \tag{2.8.3}$$

2.9 *The uniform distribution of deaths*

The assumption is frequently made that deaths between age x and age $x+1$ are uniformly distributed over the year of age. This assumption is usually very accurate for human populations (except perhaps at the very young ages), and it is a very convenient one for numerical work. It is equivalent to assuming that l_{x+t} is linear in the range $0 \leqslant t \leqslant 1$. One can deduce that

$$L_x = \tfrac{1}{2}(l_x + l_{x+1}), \tag{2.9.1}$$

$$m_x = 2q_x/(2-q_x), \tag{2.9.2}$$

$$q_x = 2m_x/(2+m_x), \tag{2.9.3}$$

and

$$_tq_x = tq_x \quad (0 \leqslant t \leqslant 1), \tag{2.9.4}$$

under this assumption.

2.10 *The concept of a stationary population*

Demographers sometimes make use of the theoretical concept of a stationary population. Some of the results can be used as approximations with real populations, but the formulae are likely to find wider use in countries achieving zero population growth.

We consider a closed population (i.e. one not subject to migration) maintained in a stationary state by a constant number Kl_0 of births per annum spread uniformly over the year. At time t, the number of persons aged between x and $x+\mathrm{d}x$ is equal to the number of survivors of those born x years ago between times $t-x-\mathrm{d}x$ and $t-x$, or ${}_xp_0 Kl_0\,\mathrm{d}x$. But ${}_xp_0 = l_x/l_0$, and we conclude that the number of persons aged between x and $x+\mathrm{d}x$ at any point of time t is

$$Kl_x\,\mathrm{d}x. \tag{2.10.1}$$

The number of persons aged x last birthday at any point of time will therefore be

$$KL_x = \int_0^1 Kl_{x+u}\,\mathrm{d}u. \tag{2.10.2}$$

To determine the number of persons passing through exact age x in a given year t_0 to t_0+1, we consider the number of persons in the age range $x-\mathrm{d}t$ to x at time t $(t_0 \leqslant t \leqslant t_0+1)$. All of these lives must reach age x, and it follows that the number of persons passing through exact age x (i.e. celebrating their xth birthday) is

$$Kl_x = \int_{t_0}^{t_0+1} Kl_x\,\mathrm{d}t. \tag{2.10.3}$$

The number of persons aged x last birthday dying in a given year can be obtained by considering the persons aged between $x+u-\mathrm{d}u$ and $x+u$ at time $t(t_0 \leqslant t \leqslant t_0+1)$. For each life the probability of death in time element $\mathrm{d}t$ is $\mu_{x+u}\,\mathrm{d}t$, and consequently the number of deaths aged x last birthday is

$$Kd_x = \int_{t_0}^{t_0+1}\int_0^1 Kl_{x+u}\mu_{x+u}\,\mathrm{d}u\,\mathrm{d}t. \tag{2.10.4}$$

According to formulae (2.10.2) and (2.8.2), the size of the stationary population is

$$Kl_0\mathring{e}_0 = K\int_0^\infty l_x\,\mathrm{d}x, \tag{2.10.5}$$

and it follows that the birth rate (and the death rate) in the population is

$$\frac{Kl_0}{Kl_0\mathring{e}_0} = \frac{1}{\mathring{e}_0}. \tag{2.10.6}$$

These techniques have been described in terms of a closed population, but they are of more general application. Consider, for example, a population maintained in a stationary state by 100,000 births and 50,000 immigrants aged exactly 20 each year. What proportion of the population is aged under 15? The size of the population is $100,000\, \mathring{e}_0 + 50,000\, \mathring{e}_{20}$, and the number under age 15 is

$$100,000\,(\mathring{e}_0 - {}_{15}p_0\, \mathring{e}_{15}).$$

The proportion under 15 can then be evaluated.

2.11 *The multiple-decrement table**

The life table described in the earlier part of this chapter has only one mode of decrement – death. The multiple-decrement table is a natural extension of this model, and it finds many applications in demographic analysis. We shall describe the techniques involved in terms of a double-decrement table, but the methods are quite general.

Let us consider a group of lives subject to two different modes of death α and β. The survivorship function[1] associated with this group of lives will be denoted by $(al)_x$. A life aged x will survive from age x to age $x+t$ with probability $(al)_{x+t}/(al)_x$, and this ratio is denoted by ${}_t(ap)_x$. The probability that a life now aged x will survive from age x to age $x+t$ and then die in the element of age dt by mode α is assumed to be

$$\frac{(al)_{x+t}}{(al)_x}(a\mu)^{\alpha}_{x+t}\mathrm{d}t. \tag{2.11.1}$$

A similar formula is true for mode β. The probability that a life now aged x will die by either mode between ages $x+t$ and $x+t+\mathrm{d}t$ is

$$\frac{(al)_{x+t}}{(al)_x}(a\mu)_{x+t}\mathrm{d}t, \tag{2.11.2}$$

and it is clear therefore that the total force of decrement

$$(a\mu)_x = (a\mu)^{\alpha}_x + (a\mu)^{\beta}_x. \tag{2.11.3}$$

We shall be interested in the number $(ad)^{\alpha}_x$ of α-deaths aged x last birthday. Clearly,

$$(ad)^{\alpha}_x = \int_0^1 (al)_{x+t}\,(a\mu)^{\alpha}_{x+t}\mathrm{d}t, \tag{2.11.4}$$

[1] There is no agreed International notation, and we shall use symbols commonly accepted in the United Kingdom and Australia. The usual North American notation is described by C. W. Jordan (1967).

and the total number of deaths aged x last birthday

$$(ad)_x = (ad)_x^\alpha + (ad)_x^\beta. \tag{2.11.5}$$

Returning to the $(al)_x$ function, it may be noted that

$$(al)_{x+1} = (al)_x - (ad)_x = (al)_x - (ad)_x^\alpha - (ad)_x^\beta. \tag{2.11.6}$$

These quantities appear in the double-decrement table 2.11.1.

TABLE 2.11.1. *A double-decrement table*

Age x	Survivorship function $(al)_x$	Deaths by mode α $(ad)_x^\alpha$	Deaths by mode β $(ad)_x^\beta$	Total deaths $(ad)_x$
50	100,000	1,151	2,401	3,552
51	96,448	1,230	2,520	3,750
52	92,698	1,308	2,646	3,954
53	88,744	1,387	2,773	4,160
54	84,584	1,462	2,897	4,359
55	80,225	1,533	3,012	4,545
56	75,680	1,596	3,116	4,712
57	70,968	1,651	3,203	4,854
58	66,114	1,694	3,271	4,965
59	61,149	1,723	3,316	5,039

The probability that a life aged x will die by mode α before attaining age $x+1$ is denoted by

$$(aq)_x^\alpha = (ad)_x^\alpha/(al)_x, \tag{2.11.7}$$

and because the probability of death by either mode during that year of life is

$$(aq)_x = (ad)_x/(al)_x, \tag{2.11.8}$$

we conclude that

$$(aq)_x = (aq)_x^\alpha + (aq)_x^\beta. \tag{2.11.9}$$

The quantities $(aq)_x^\alpha$ and $(aq)_x^\beta$ are sometimes referred to as *dependent rates of decrement.*

Most of the quantities associated with multiple-decrement tables have now been introduced. It is to be noted that our approach to the multiple-decrement table differs from our development of the life table; instead of deriving the forces of decrement from the other multiple-decrement functions, we have used the forces from the outset to develop the theory. No equation like formula (2.3.2) has yet been derived, and this is our next task.

For the total force of decrement, there is no problem:

$$(a\mu)_x = \frac{-1}{(al)_x}\frac{d}{dx}(al)_x, \tag{2.11.10}$$

and
$$_t(ap)_x = \exp\left\{-\int_0^t (a\mu)_{x+u}\,du\right\}. \tag{2.11.11}$$

To derive a relationship for $(a\mu)_x^\alpha$, it is convenient to define

$$(al)_x^\alpha = \sum_{t=0}^{\infty} (ad)_{x+t}^\alpha. \tag{2.11.12}$$

Then by definition,

$$(a\mu)_x^\alpha = \lim_{h\to 0} \frac{(al)_x^\alpha - (al)_{x+h}^\alpha}{h(al)_x}$$

$$= \frac{-1}{(al)_x}\frac{d}{dx}(al)_x^\alpha, \tag{2.11.13}$$

and it is clear that equation (2.11.3) is valid. Note the omission of the superscript α in the denominator of equation (2.11.13).

The letter a in $(al)_x$ is arbitrary. If we are considering two different multiple-decrement tables, it might be convenient to use the letter b (or any other letter) for the second table, and define quantities like $(bl)_x$, $(b\mu)_x$, etc.

2.12 *The multiple-decrement table and its related single-decrement tables**

It is possible to imagine a population in which the mode of death β is non-existent, but deaths still occur by mode α. For this population, the survivorship function will be denoted by l_x^α and a life aged x will survive from age x to age $x+t$ with probability $_tp_x^\alpha = l_{x+t}^\alpha / l_x^\alpha$.

The probability that a life now aged x will survive from age x to age $x+t$ and then die in the element of age dt is

$$\frac{l_{x+t}^\alpha}{l_x^\alpha}\mu_{x+t}^\alpha\,dt. \tag{2.12.1}$$

This result was proved in section 2.3. The population may be analysed using the ordinary life-table techniques and life-table symbols will be used with a superscript α to indicate this particular population. One formula which we shall need is the following:

$$_tp_x^\alpha = \exp\left(-\int_0^t \mu_{x+u}^\alpha\,du\right), \tag{2.12.2}$$

which is readily proved using equation (2.3.2).

It is also possible to imagine a population with a single mode of decrement β, and this population can be analysed using the life-table methods. A superscript β will be used for this population.

Let us now arrange for the forces of decrement μ_x^α and μ_x^β to be equal to $(a\mu)_x^\alpha$ and $(a\mu)_x^\beta$ respectively. Then the single-decrement tables characterized by l_x^α and l_x^β are called the *related single-decrement tables* of the double-decrement table $(al)_x$. We shall be interested in relationships between these tables.

The starting point is a comparison of equations (2.11.11) and (2.12.2). We see that

$$_t(ap)_x = {}_tp_x^\alpha\, {}_tp_x^\beta, \tag{2.12.3}$$

and it is worth noting that this equation implies the independence of the two survival probabilities on the right-hand side. If we know the single-decrement functions l_x^α and l_x^β, we can compute the $(al)_x$ column using equation (2.12.3) and then deduce the values of $(ad)_x$.

To determine values of $(ad)_x^\alpha$ and $(ad)_x^\beta$, some further theory is necessary, and we shall make the *assumption* that decrements in the related single-decrement tables are uniformly distributed over the year of life[2] (section 2.9). Then for $0 \leqslant t \leqslant 1$,

$$_tq_x^\alpha = tq_x^\alpha. \tag{2.12.4}$$

Furthermore, we know that

$$_tq_x^\alpha = \int_0^t {}_up_x^\alpha \mu_{x+u}^\alpha \,\mathrm{d}u. \tag{2.12.5}$$

But according to equation (2.12.4), $_tq_x^\alpha$ is proportional to t for $0 \leqslant t \leqslant 1$, and it follows that

$$_tp_x^\alpha \mu_{x+t}^\alpha = q_x^\alpha \quad (0 \leqslant t \leqslant 1). \tag{2.12.6}$$

From equations (2.11.4) and (2.11.7), we can see that

$$\begin{aligned}(aq)_x^\alpha &= \int_0^1 {}_t(ap)_x\,(a\mu)_{x+t}^\alpha \,\mathrm{d}t \\ &= \int_0^1 {}_tp_x^\alpha\, {}_tp_x^\beta\, \mu_{x+t}^\alpha \,\mathrm{d}t \\ &= \int_0^1 (1 - {}_tq_x^\beta)\,({}_tp_x^\alpha\, \mu_{x+t}^\alpha)\,\mathrm{d}t \\ &= \int_0^1 (1 - tq_x^\beta)\, q_x^\alpha \,\mathrm{d}t \\ &= q_x^\alpha(1 - \tfrac{1}{2}q_x^\beta),\end{aligned} \tag{2.12.7}$$

[2] A different assumption is made in question 28 of section 2.14.

and in a similar manner,

$$(aq)_x^\beta = q_x^\beta(1-\tfrac{1}{2}q_x^\alpha). \tag{2.12.8}$$

The quantities q_x^α and q_x^β are often referred to as the *independent rates of decrement*, because they refer to the related single-decrement tables, and they do not depend on any other decrement. (The *dependent* rates are functions of quantities from both single-decrement tables.)

It is clear from equations (2.12.7) and (2.12.8) that *if* the decrements are uniform in the related single-decrement tables, we can readily compute all the items in the double-decrement table using the independent rates q_x^α and q_x^β. The inverse problem of determining q_x^α and q_x^β from $(aq)_x^\alpha$ and $(aq)_x^\beta$ is a little more complicated. One approach is to eliminate q_x^β from equations (2.12.7) and (2.12.8). We obtain a quadratic equation for q_x^α, and the coefficients are functions of $(aq)_x^\alpha$ and $(aq)_x^\beta$.

Very often the rates are fairly small, and it is simpler to employ an iterative method based on the equations

$$\left.\begin{aligned} q_x^\alpha &= (aq)_x^\alpha(1-\tfrac{1}{2}q_x^\beta)^{-1}; \\ \text{and} \qquad q_x^\beta &= (aq)_x^\beta(1-\tfrac{1}{2}q_x^\alpha)^{-1}. \end{aligned}\right\} \tag{2.12.9}$$

The quantities $(aq)_x^\alpha$ and $(aq)_x^\beta$ are used as initial values for q_x^α and q_x^β respectively.

2.13 *Examples*

Example 1.[3] Prove that

$$\mu_x \doteqdot -\tfrac{1}{2}(\log p_{x-1}+\log p_x). \tag{2.13.1}$$

Solution. By definition

$$\mu_{x+t} = -\frac{\mathrm{d}}{\mathrm{d}t}\log l_{x+t}.$$

It follows that

$$\begin{aligned}\int_{-1}^{1}\mu_{x+t}\,\mathrm{d}t &= \log l_{x-1}-\log l_{x+1} \\ &= \log l_{x-1}-\log l_x+\log l_x-\log l_{x+1} \\ &= -\log p_{x-1}-\log p_x.\end{aligned}$$

If μ_{x+t} is approximately linear in the range $-1 \leqslant t \leqslant 1$, the integral may be approximated by $2\mu_x$. Result (2.13.1) follows immediately.

[3] The logarithms in this example and throughout the whole book are natural logarithms (to base e).

Example 2.[4] Show that if $\mu_x = Bc^x$, where B and c are constants, then

$$l_x = kg^{c^x}$$

where k and g are constants.

Solution. By definition

$$\mu_x = -\frac{\mathrm{d}}{\mathrm{d}x}\log l_x = Bc^x.$$

It follows that

$$\log l_x = -\int B\mathrm{e}^{x\log c}\mathrm{d}x$$

$$= -\frac{B}{\log c}c^x + \text{constant},$$

whence $l_x = kg^{c^x}$,

where $g = \exp\left(\frac{-B}{\log c}\right)$.

Example 3. Consider a table $(al)_x$ with two decrements α and β. The decrement α is distributed uniformly over the year of age and the decrement β is concentrated at age $x+k$, where $0 \leqslant k < 1$. Derive relationships between the dependent and independent rates of decrement.

Solution. For decrement α, equation (2.12.6) is still valid, and we note that for decrement β,

$${}_tp_x^\beta = 1 \quad (0 \leqslant t < k);$$

and $${}_tp_x^\beta = 1 - q_x^\beta \quad (k \leqslant t < 1).$$

Then $$(aq)_x^\alpha = \int_0^1 {}_tp_x^\beta\, {}_tp_x^\alpha\, \mu_{x+t}^\alpha\,\mathrm{d}t$$

$$= \int_0^k {}_tp_x^\alpha\, \mu_{x+t}^\alpha\,\mathrm{d}t + \int_k^1 (1 - q_x^\beta)\, {}_tp_x^\alpha\, \mu_{x+t}^\alpha\,\mathrm{d}t$$

$$= kq_x^\alpha + (1-k)(1-q_x^\beta)\,q_x^\alpha$$

$$= q_x^\alpha\{1-(1-k)\,q_x^\beta\}.$$

[4] This 'law' of mortality was suggested by Benjamin Gompertz in 1825. A modification of the curve was suggested some years later by W. M. Makeham (1860): $\mu_x = A+Bc^x$. Neither 'law' is applicable to the juvenile ages.

But $$(ap)_x = p_x^\alpha p_x^\beta$$
and we deduce that
$$(aq)_x^\beta = q_x^\beta(1-kq_x^\alpha).$$
The special case $k = \frac{1}{2}$ should be noted.

2.14 *Exercises*

1 On the basis of the Australian life table (males) 1961, find the probability that a man aged 30 will
(*a*) survive to age 40;
(*b*) die before reaching age 50;
(*c*) die in his 50th year (i.e. between ages 49 and 50);
(*d*) die between ages 40 and 50.

2 Show that an increase in the constant B in Makeham's Law $\mu_x = A+Bc^x$ is equivalent to a constant increase in age.

3 (i) Find q_{50} given that $\mu_{50} = 0.01098$ and $\mu_{51} = 0.01173$.
(ii) Find μ_{40} given that $p_{39} = 0.99469$ and $p_{40} = 0.99438$.
(iii) Find μ_{40} given that $l_{38} = 81{,}778$
$$l_{39} = 81{,}367$$
$$l_{40} = 80{,}935$$
$$l_{41} = 80{,}480$$
$$l_{42} = 79{,}999.$$
Derive any formulae you use.

4 Is it possible for μ_x to be greater than one?

5 For a certain life table $l_x = 20{,}900-80x-x^2$.
(i) What is the ultimate age in the life table?
(ii) Find μ_x and q_x.
(iii) Find ${}_{10}p_{20}$.

6 You are given that
$$\mu_x = \frac{1}{(a_0+a_1x)\,(b_0+b_1x)}.$$
Find an expression for l_x.

7 The force of mortality for a particular population follows Makeham's Law, and $\mu_x = A+Bc^x$. Prove that
$$l_x = ks^x g^{c^x},$$
where k, s and g are constants.

8 (i) What conditions must a survivorship function l_x satisfy?
(ii) Show that $1-x/106$ satisfies these conditions.
(iii) What is the ultimate age in this table?
(iv) What is the formula for μ_x?
(v) What is the probability that a life aged 13 will die before age 43?

9 Prove the following results:

(*a*) $$\frac{e_x \cdot e_{x+1} \cdots e_{x+n-1}}{(1+e_{x+1})(1+e_{x+2}) \cdots (1+e_{x+n})} = {}_np_x.$$

(*b*) $({}_{\frac{1}{2}}p_x + {}_{\frac{3}{2}}p_x + {}_{\frac{5}{2}}p_x + \ldots) - ({}_1p_x + {}_2p_x + {}_3p_x + \ldots) \doteqdot \frac{1}{2}.$

10 Prove that the complete expectation of life at age x is T_x/l_x where

$$T_x = \int_0^\infty l_{x+t}\,\mathrm{d}t.$$

11 Prove that the average age at death of those persons who die between age x and age y is

$$x + \{T_x - T_y - (y-x)\, l_y\}/(l_x - l_y).$$

12 For an animal population, the force of mortality is independent of age and equal to μ. Find
(i) the number of deaths between ages x and $x+\mathrm{d}x$ out of l_0 births;
(ii) l_x;
(iii) the average age at death; and
(iv) $\mathring{e}_x$.

13 Determine the first partial derivatives of ${}_tp_x$ with respect to x and with respect to t. Evaluate these derivatives using the Australian life table (males) 1961 for $x = 60$ and $t = 5$.

14 Two lives now aged x and y respectively are subject to Gompertz mortality from the same table (i.e. $\mu_x = Bc^x$ and $\mu_y = Bc^y$). What is the probability that the life aged x dies before the life aged y?

15 Generalize the result of question 14 for the case of three or more lives.

16 Prove that

$$\mu_x = \left(1 + \frac{\mathrm{d}}{\mathrm{d}x}\mathring{e}_x\right) \Big/ \mathring{e}_x.$$

Check this result numerically at age 60 using the Australian life table (males) 1961.

17 For a certain population $l_x = 10{,}000(121-x)^{\frac{1}{2}}$. Find
(i) μ_x;
(ii) q_x; and
(iii) the probability that a life aged 0 will die between ages 21 and 40.

18 Prove that

$$m_x = -\frac{1}{L_x}\frac{\mathrm{d}}{\mathrm{d}x}L_x.$$

Using this result, prove that $m_x \doteqdot \mu_{x+\frac{1}{2}}$.

19 Two lives A and B are now aged 65 and 75 respectively. Both are Australian males. Write down an integral for the probability that B dies before A and within 20 years. Evaluate this probability using the repeated Simpson rule of integration and an interval of 5 years.

20 For a certain life table, $l_x = a+bx+cx^2$, when $x_1 \leqslant x \leqslant x_2$. Find the value of x in this range for which the error in l_x, introduced by the assumption of a uniform distribution of deaths for $x_1 \leqslant x \leqslant x_2$, is a maximum.

21 A new mortality table is prepared from a standard mortality table by doubling the force of mortality at all ages. Is the mortality rate q_x' at any given age under the new table less than double, exactly double, or more than double the mortality rate q_x of the standard table?

22 In a certain life table,

$$\mu_x = 0.15-0.10x \quad \text{for} \quad 0 \leqslant x \leqslant \tfrac{1}{2},$$

and

$$\mu_x = (0.01)^x \quad \text{for} \quad \tfrac{1}{2} \leqslant x \leqslant 1.$$

Find l_1 assuming that $l_0 = 100{,}000$.

23 Of 21,500 persons who reach age 72, 11 die within a month, 16 die in the second month, 22 die in the third month, 30 die in the fourth month, 41 die in the fifth month, and so on. Determine,

(i) the number of years of life lived by these persons between ages 72 and 73;
(ii) the death rate at age 72;
(iii) the central death rate at age 72; and
(iv) the force of mortality at age $72\tfrac{1}{2}$.

24 (i) In a certain large population there are $P_x(t)$ persons aged x last birthday at time t. Write down an integral in terms of $P_x(t)$ for the number of years of life lived between exact age x and exact age $x+1$ during the time interval 0 to 1 years.
(ii) During the time interval (0, 1) there are θ_x deaths aged x last birthday. When θ_x is divided by the integral obtained in part (i), the result is an estimate of a standard mortality function. Which one?
(iii) Given the following data, construct a table of l_x, d_x, p_x, q_x, L_x and m_x for ages 41 to 45, assuming that $l_{41} = 10{,}000$.

Age x	Population aged x last birthday in the middle of 1968	Deaths aged x last birthday during 1968
41	3,102	25
42	2,196	18
43	2,617	22
44	2,595	22
45	3,002	26

Hint for part (iii). Use the results of (i) and (ii), and make use of the approximation

$$\int_0^1 f(t)\,dt \doteqdot f(\tfrac{1}{2}).$$

25 The present discounted value of \$1 payable in t years' time is v^t where $v = (1+i)^{-1}$ and i denotes the annual rate of interest. Write down an integral representing the expected discounted value of \$1 payable at the instant of death of a life now aged x.

26 Determine the variance of the discounted assurance in question 25.

27 Use the data in table 2.11.1 to evaluate q_x^α and q_x^β for ages 50 to 59. Also evaluate μ_x^α and μ_x^β in this range.

28 A multiple-decrement table has two modes of decrement α and β. The total decrement is distributed uniformly over the year of age, and so is the decrement in the independent single decrement table α. Prove that

$$_tp_x^\beta \doteqdot 1 - tq_x^\beta(1-q_x^\alpha) - t^2q_x^\alpha q_x^\beta(1-q_x^\alpha)$$

and deduce that

$$(aq)_x^\alpha \doteqdot q_x^\alpha(1-\tfrac{1}{2}q_x^\beta + \tfrac{1}{6}q_x^\alpha q_x^\beta).$$

29 The quality control unit of a light-bulb factory has found that their product has a life-time (in weeks) with the probability density

$$(a+2bx)\exp\{-(bx^2+ax)\} \quad (x \geqslant 0),$$

where a and b are positive constants. The factory supplies light-bulbs to an underground railway system where they are put into continuous use until they either fail or are stolen. The force of thieving is approximately constant over time and equal to τ per week.
(i) Determine the total force of decrement $(a\mu)_x$ for a light-bulb that has been installed in the underground x weeks.
(ii) Determine the distribution of the light-bulb life-time for a newly-installed light-bulb.
(iii) Determine $_t(ap)_x$ for these underground light-bulbs.

30 Prove the following formula which is applicable to a multiple-decrement table

$$(a\mu)_x^\alpha \doteqdot \frac{8\{(ad)_{x-1}^\alpha + (ad)_x^\alpha\} - \{(ad)_{x-2}^\alpha + (ad)_{x-1}^\alpha + (ad)_x^\alpha + (ad)_{x+1}^\alpha\}}{12(al)_x}.$$

3
The deterministic population models of T. Malthus, A. J. Lotka, and F. R. Sharpe and A. J. Lotka

3.1 *Introduction*

Thomas Malthus can probably be credited with formulating the first population mathematical model in 1798. The following quotations come from his *Essay on the Principle of Population.*

> 'I think I may fairly make two postulata
>
> First, That food is necessary to the existence of man. (*page* 11)
>
> Secondly, That the passion between the sexes is necessary, and will remain nearly in its present state.
>
> ...
>
> Assuming then, my postulata as granted, I say, that the power of population is indefinitely greater than the power in the earth to produce subsistence for man.
>
> Population, when unchecked, increases in a geometrical ratio. Subsistence increases only in an arithmetical ratio. (*page* 13) A slight acquaintance with numbers will shew the immensity of the first power in comparison of the second.
>
> ...
>
> In the United States of America, where the means of subsistence have been more ample, the manners of the people more pure,[1] and consequently the checks to early marriage fewer, than in (*page* 20) any of the modern states of Europe, the population has been found to double itself in twenty-five years.'

[1] Thomas Malthus, a Church of England clergyman, regarded the use of contraceptive devices as a vice.

If we denote the population size by n, time by t and the rate of increase of the population by r, Malthus's model is represented by the following first-order differential equation

$$\frac{dn}{dt} = rn. \tag{3.1.1}$$

The mathematical model of A. J. Lotka (1907) was based on this differential equation. Many of the models discussed in this book are such that the total populations grow exponentially asymptotically, and they then obey equation (3.1.1).

Usually, a population cannot grow exponentially for ever. If the growth rate r is negative, it will disappear. If the growth rate is positive, it will become too large for the environment to support it. Population mathematicians have therefore modified equation (3.1.1) to obtain the differential equation for the so-called *logistic population*:

$$\frac{dn}{dt} = rn\left(1 - \frac{n}{N}\right). \tag{3.1.2}$$

This type of population grows exponentially while it is small, but the growth rate tapers off as the population size increases, and the population cannot exceed a certain maximum size N (unless of course it had been made greater than N to begin with artificially, in which case it will decrease to size N). For human populations, the logistic law of growth has not proved very satisfactory, and we shall only mention it once again in this book.

3.2 *The continuous-time model of Sharpe and Lotka*

Although T. Malthus (1798) and A. J. Lotka (1907) proposed elementary mathematical models for human populations, the paper which might be considered to be the beginning of the subject of Population Mathematics is that of F. R. Sharpe and A. J. Lotka in 1911.

They considered the male population alone, and assumed that the growth of the female population was such as to justify the assumptions of constant fertility and mortality rates. The female sex has a shorter reproductive life-span, and illegitimate births are more readily attributable to the mother. Demographers therefore usually find it more convenient to apply this one-sex model to the female sex, and we shall describe the model in terms of females. It can of course be applied to either sex.

We define $F(x, t)\,dx$ to be the number of females at time t whose ages lie between x and $x+\mathrm{d}x$, and $B(t)\,\mathrm{d}t$ to be the number of female births during the time element $(t, t+\mathrm{d}t)$. From these definitions, it is clear that

$$\left.\begin{aligned} F(x,t)\,\mathrm{d}x &= B(t-x)\,{}_x p_0\,\mathrm{d}x \quad (0 \leqslant x \leqslant t);\\ \text{and}\quad F(x,t)\,\mathrm{d}x &= F(x-t,0)\,{}_t p_{x-t}\,\mathrm{d}x \quad (x > t).\end{aligned}\right\} \tag{3.2.1}$$

Let the number of female births in time element $\mathrm{d}t$ to the $F(x,t)\,\mathrm{d}x$ females whose ages lie between x and $x+\mathrm{d}x$ be $F(x,t)\,\mathrm{d}x\,\lambda(x)\,\mathrm{d}t$. Then clearly,

$$\begin{aligned} B(t) &= \int_0^\infty F(x,t)\,\lambda(x)\,\mathrm{d}x\\ &= \int_0^t B(t-x)\,{}_x p_0\,\lambda(x)\,\mathrm{d}x + \int_0^\infty F(x,0)\,{}_t p_x\,\lambda(x+t)\,\mathrm{d}x.\end{aligned}$$

That is

$$B(t) = \int_0^t B(t-x)\,{}_x p_0\,\lambda(x)\,\mathrm{d}x + G(t), \tag{3.2.2}$$

where

$$G(t) = \int_0^\infty F(x,0)\,{}_t p_x\,\lambda(x+t)\,\mathrm{d}x. \tag{3.2.3}$$

This is the basic integral equation for the Sharpe and Lotka one-sex deterministic population model.

Let us follow the solution of E. C. Rhodes (1940). Reproduction takes place over only part of the life span of a female, between ages α and β say. Therefore

$$G(t) = 0 \quad (t > \beta). \tag{3.2.4}$$

Consider $t > \beta$, and make a trial solution

$$B(t) = A\,\mathrm{e}^{rt}. \tag{3.2.5}$$

When this is substituted in equation (3.2.2), we obtain

$$A\,\mathrm{e}^{rt} = A\,\mathrm{e}^{rt}\int_\alpha^\beta \mathrm{e}^{-rx}\,{}_x p_0\,\lambda(x)\,\mathrm{d}x.$$

Then r is the solution of the integral equation

$$\int_\alpha^\beta e^{-rx}\,{}_x p_0\,\lambda(x)\,\mathrm{d}x = 1. \tag{3.2.6}$$

Theorem 3.2.1. The integral equation (3.2.6) has exactly one real solution $r = r_0$. Any complex roots $\{r_j\}$ occur in complex conjugate pairs, and $r_0 > \mathrm{Real}\,(r_j)$.

Proof. (i) To prove that there is only one real root r_0, define

$$f(r) = \int_{\alpha}^{\beta} e^{-rx} {}_xp_0 \lambda(x) \, dx. \tag{3.2.7}$$

Then
$$\frac{d}{dr} f(r) = -\int_{\alpha}^{\beta} x e^{-rx} {}_xp_0 \lambda(x) \, dx. \tag{3.2.8}$$

Clearly, $f(r) \to \infty$ as $r \to -\infty$, and $f(r) \to 0$ as $r \to \infty$. But $f(r)$ is a continuous function of r. Hence, at least one real solution for equation (3.2.6) exists.

The functions x, e^{-rx}, ${}_xp_0$ and $\lambda(x)$ are all non-negative over the range of integration in (3.2.8), and they are all simultaneously positive over part of the range of integration. Hence $df(r)/dr$ is negative for all real r. It follows that $f(r)$ is monotone strictly decreasing, and hence there is only one real solution $r = r_0$ of the equation $f(r) = 1$.

(ii) Consider now the complex roots, and suppose that $u + iv$ is one such root. Equation (3.2.6) becomes

$$\int_{\alpha}^{\beta} e^{-ux} \{\cos(-vx) + i \sin(-vx)\} {}_xp_0 \lambda(x) \, dx = 1.$$

Equating real and imaginary parts,

$$\left.\begin{aligned} \int_{\alpha}^{\beta} e^{-ux} \cos(vx) \, {}_xp_0 \lambda(x) \, dx &= 1, \\ \text{and} \quad \int_{\alpha}^{\beta} e^{-ux} \sin(vx) \, {}_xp_0 \lambda(x) \, dx &= 0. \end{aligned}\right\} \tag{3.2.9}$$

It follows that $u - iv$ is also a complex root of equation (3.2.6), and the complex roots occur in conjugate pairs.

Because $\cos(vx) < 1$ for some values of x in the range of integration for equation (3.2.9),

$$\int_{\alpha}^{\beta} e^{-ux} \, {}_xp_0 \lambda(x) \, dx > 1.$$

But
$$\int_{\alpha}^{\beta} e^{-r_0 x} \, {}_xp_0 \lambda(x) \, dx = 1.$$

Therefore $u < r_0$. That is,

$$r_0 > \text{Real}(r_j). \tag{3.2.10}$$

This completes the proof of the theorem.

If r_i $(i = 0, 1, 2, \ldots)$ is a root of equation (3.2.6), then

$$B(t) = A_i e^{r_i t} \quad (i = 0, 1, 2, \ldots) \tag{3.2.11}$$

is a solution of equation (3.2.2) for $t > \beta$, but there is no guarantee that *all* the roots of equation (3.2.2) are of this form.

Equation (3.2.6) might for example have a multiple complex root. Consider one such root r_c of multiplicity 2. Then

$$\left.\begin{aligned} f(r_c) - 1 &= 0, \\ \text{and} \quad f'(r_c) &= 0. \end{aligned}\right\} \tag{3.2.12}$$

The second equation means that

$$\int_\alpha^\beta x\,e^{-r_c x}\,{}_xp_0\,\lambda(x)\,dx = 0$$

and we find that equation (3.2.2) has a solution (for $t > \beta$) of the form

$$B(t) = (C_0 + C_1 t)\,e^{r_c t}. \tag{3.2.13}$$

The whole problem of solutions to equation (3.2.2) which are not of the form e^{rt} centres around the existence or non-existence of multiple complex roots for equation (3.2.6). These are not found empirically, although examples may be constructed. A rigorous, but rather complicated analysis has been given by W. Feller (1941), and the interested reader should examine this reference. The problem recurs in the discrete analysis of chapter 4, and a rigorous solution is given there for the discrete case (section 4.5).

As t increases, the behaviour of $B(t)$ is dominated by the real root r_0, which is of multiplicity 1. To prove this, consider first the case in which all the roots of equation (3.2.6) are distinct. Then

$$B(t) = \sum_j A_j e^{r_j t} = A_0 e^{r_0 t} + \sum_{j \neq 0} A_j e^{u_j t}\{\cos(v_j t) + i \sin(v_j t)\}.$$

Because $r_0 > u_j$, $e^{u_j t}$ is negligible compared with $e^{r_0 t}$ for large t, and

$$B(t) \simeq A_0 e^{r_0 t}. \tag{3.2.14}$$

Roots of finite multiplicity cause no difficulty because

$$(C_0 + C_1 t + \ldots + C_{n-1} t^{n-1})\,e^{u_j t}$$

is negligible compared with $e^{r_0 t}$ for large t.

3.3 *The stable age distribution*

According to the definition in section 3.2, the number of females at time t in the age range $(x, x+dx)$ is $F(x,t)\,dx$. Using equations (3.2.1) and (3.2.14), we conclude that

$$F(x,t) \simeq {}_xp_0\,A_0\,e^{r_0(t-x)}. \tag{3.3.1}$$

The proportion of females in the age group $(x, x+\mathrm{d}x)$ is therefore given by

$$\frac{F(x,t)\,\mathrm{d}x}{\int_0^\infty F(u,t)\,\mathrm{d}u} \simeq \frac{{}_xp_0\,\mathrm{e}^{-r_0x}\,\mathrm{d}x}{\int_0^\infty {}_up_0\,\mathrm{e}^{-r_0u}\,\mathrm{d}u}. \tag{3.3.2}$$

This proportion is independent of t for large t, so a stable asymptotic age distribution exists (as was known to Euler (footnote 3, page 1)), and that distribution is given by equation (3.3.2).

3.4 *Numerical values of r_0*

Consider equation (3.2.6), and define

$$R_n = \int_\alpha^\beta x^n\,{}_xp_0\,\lambda(x)\,\mathrm{d}x \quad (n = 0, 1, 2, \ldots). \tag{3.4.1}$$

R_0 is the average number of daughters that will be born to a female now aged 0, and it is known as the *net reproduction rate* (NRR). In general, R_0 is not equal to unity, and it is convenient to divide both sides of equation (3.2.6) by R_0:

$$\int_\alpha^\beta \frac{\mathrm{e}^{-rx}\,{}_xp_0\,\lambda(x)}{R_0}\,\mathrm{d}x = \frac{1}{R_0}. \tag{3.4.2}$$

It is immediately apparent that $\{{}_xp_0\,\lambda(x)\}/R_0$ is a probability-density function in the range (α, β), and it follows that the left-hand side of equation (3.4.2) is a moment-generating function $M(-r)$. Taking logarithms of both sides, we obtain

$$K(-r) = -\log R_0 + 2n\pi\mathrm{i} \quad (n = 0, \pm 1, \pm 2, \ldots), \tag{3.4.3}$$

and $K(-r)$ is the cumulant-generating function of the net maternity function of a stationary population.

The distribution is over the finite portion of the positive axis between α and β, so all the moments exist, and they are all finite. It follows that all the cumulants exist and they are all finite. The cumulant-generating function[2] may be expanded in an infinite series (M. G. Kendall and A. Stuart, 1963) and we obtain the equation:

$$r\kappa_1 - \frac{1}{2!}r^2\kappa_2 + \frac{1}{3!}r^3\kappa_3 - \ldots = \log R_0 - 2n\pi\mathrm{i}. \tag{3.4.4}$$

This equation with $n = 0$ is the one we now need to solve to evaluate r_0.

A. J. Lotka noted that r_0 is small. An approximate value of r_0 is therefore given by the solution of the quadratic equation:

$$\tfrac{1}{2}r^2\kappa_2 - r\kappa_1 + \log R_0 = 0. \tag{3.4.5}$$

[2] The first cumulant κ_1 is the average age of childbearing in the population.

We are now faced with a paradox. Equation (3.4.5) has two real[3] roots. Yet theorem 3.2.1 tells us that there is only one real solution to equation (3.2.6).

Equation (3.4.5) is identical with equation (3.4.4) when $n = 0$ provided all the cumulants above the second are zero. The net maternity function $\{_x p_0 \lambda(x)\}/R_0$ is then normal over the range $(-\infty, \infty)$. The use of equation (3.4.5) is therefore equivalent to the assumption that the net maternity function is normal. But part of the normal density curve *must* lie above the negative portion of the age-axis, and it is easy to deduce that this is the reason for the second real root. It is also easy to see that the smaller root of equation (3.4.5) must be taken as the approximate real solution to equation (3.4.4). Thus

$$r_0 \doteqdot \{\kappa_1 - (\kappa_1^2 - 2\kappa_2 \log R_0)^{\frac{1}{2}}\}/\kappa_2. \tag{3.4.6}$$

Many other methods for determining r_0 have been suggested. Let us use an iterative method to obtain an explicit series for r_0, and re-write equation (3.4.4) in the following form

$$r = \left(\log R_0 + \frac{1}{2!} r^2 \kappa_2 - \frac{1}{3!} r^3 \kappa_3 + \ldots\right) \Big/ \kappa_1 \tag{3.4.7}$$

or

$$r = C_0 + C_2 r^2 + C_3 r^3 + C_4 r^4 + \ldots \tag{3.4.8}$$

It is usual to select $r = 0$ as the initial value. Then after the first iteration, $r = C_0$. C_0 is very small, and we shall therefore ignore powers of C_0 above the fourth. After the fifth iteration,

$$r = C_0 + C_2 C_0^2 + (2C_2^2 + C_3) C_0^3 + (5C_2^3 + 5C_2 C_3 + C_4) C_0^4. \tag{3.4.9}$$

It may be noted that the coefficient of C_0^j remains fixed after the jth iteration, and the series converges rapidly due to the smallness of $C_0 = (\log R_0)/\kappa_1$. Formula (3.4.9) may be written in the form

$$r_0 = \left(\frac{\log R_0}{\kappa_1}\right) + \frac{1}{2}\left(\frac{\kappa_2}{\kappa_1}\right)\left(\frac{\log R_0}{\kappa_1}\right)^2 + \left\{\frac{1}{2}\left(\frac{\kappa_2}{\kappa_1}\right)^2 - \frac{1}{6}\left(\frac{\kappa_3}{\kappa_1}\right)\right\}\left(\frac{\log R_0}{\kappa_1}\right)^3 + \left\{\frac{5}{8}\left(\frac{\kappa_2}{\kappa_1}\right)^3 - \frac{5}{12}\left(\frac{\kappa_2 \kappa_3}{\kappa_1^2}\right) + \frac{1}{24}\left(\frac{\kappa_4}{\kappa_1}\right)\right\}\left(\frac{\log R_0}{\kappa_1}\right)^4 + O\left(\frac{\log R_0}{\kappa_1}\right)^5. \tag{3.4.10}$$

[3] The reader may wonder why the roots of the quadratic are necessarily real. Typically, κ_1 has a value of about 27 and κ_2 a value of about 35. A net reproduction rate R_0 greater than three would be exceptional.

This is an explicit series for r_0, and in fact it is a Taylor series. The formula is of little use computationally, but it is of theoretical interest, because it allows us to examine the effects on r_0 of changes in the cumulants.

3.5 *The effects of the cumulants on r_0*

The cumulants of the net maternity functions $\{_xp_0\lambda(x)\}/R_0$ of two populations will usually differ, and it is interesting to examine their effects on r_0. It is soon apparent from equation (3.4.10) that

$$\frac{\partial r_0}{\partial R_0} = \frac{1}{R_0\kappa_1} + \frac{\kappa_2}{\kappa_1}\frac{\log R_0}{R_0\kappa_1^2} \tag{3.5.1}$$

plus terms of smaller order, and that

$$\frac{\partial r_0}{\partial \kappa_n} = \frac{1}{n!\,\kappa_1}\left(-\frac{\log R_0}{\kappa_1}\right)^n \tag{3.5.2}$$

plus terms of smaller order. These derivatives indicate the relative importance of the various parameters.

TABLE 3.5.1. *Parameters of the net maternity function for the U.S. population**

Year	R_0	$\log R_0$	κ_1	κ_2	κ_3	κ_4	r_0
1964	1.526	0.42265	26.53	34.39	123.1	−137	0.0161
1965	1.395	0.33289	26.52	35.12	123.9	−191	0.0127

* N. Keyfitz (1968), table 5.5.

A numerical example is instructive. According to N. Keyfitz (1968), R_0 and the cumulants for the United States female population in 1964 and 1965 are those quoted in table 3.5.1. In 1964, the unique real root r_0 was 0.0161, and in 1965 it was equal to 0.0127. During that year, the first four cumulants changed very little, but R_0 changed substantially. According to formula (3.5.1), the change in r_0 was approximately

$$\Delta r_0 = \left(\frac{1}{R_0\kappa_1} + \frac{\kappa_2}{\kappa_1}\frac{\log R_0}{R_0\kappa_1^2}\right)\Delta R_0 = -0.0033.$$

The actual change in r_0 was -0.0034. Even when the second term in the above calculation is omitted, a reasonably accurate estimate for the change in r_0 is obtained: -0.0032.

3.6 *Numerical values of A_0*

If the roots of equation (3.2.6) are distinct, we have seen that

$$B(t) = \sum_{j=0}^{\infty} A_j e^{r_j t}. \tag{3.6.1}$$

The dominant root is r_0, and methods for calculating it were described in section 3.4. We now require a method for determining A_0.

Equation (3.2.2) may be written in the following form:

$$B(t) = \int_0^t B(t-x)\,\phi(x)\,dx + G(t), \tag{3.6.2}$$

where $$\phi(x) = {}_x p_0 \lambda(x). \tag{3.6.3}$$

Taking the Laplace transforms of both sides of equation (3.6.2), we have

$$B^*(r) = B^*(r)\,\phi^*(r) + G^*(r) \tag{3.6.4}$$

where
$$\left.\begin{aligned} B^*(r) &= \int_0^\infty e^{-rt} B(t)\,dt; \\ \phi^*(r) &= \int_0^\infty e^{-rt} \phi(t)\,dt; \\ \text{and}\quad G^*(r) &= \int_0^\infty e^{-rt} G(t)\,dt. \end{aligned}\right\} \tag{3.6.5}$$

It follows from equation (3.6.4) that

$$B^*(r) = \frac{G^*(r)}{1-\phi^*(r)}. \tag{3.6.6}$$

Clearly the zeros of $1-\phi^*(r)$ are the roots of equation (3.2.6). Let us assume that they are all distinct, and the right-hand side of equation (3.6.6) may be expanded as follows:

$$B^*(r) = \frac{G^*(r)}{1-\phi^*(r)} = \frac{A_0}{r-r_0} + \sum_{j=1}^{\infty} \frac{A_j}{r-r_j}. \tag{3.6.7}$$

Then $$A_0 = \lim_{r\to r_0} \frac{(r-r_0)\,G^*(r)}{1-\phi^*(r)} = \left[\frac{G^*(r)}{-d\phi^*(r)/dr}\right]_{r=r_0} \tag{3.6.8}$$

$G^*(r)$ and $\phi^*(r)$ are defined in equations (3.6.5). Both $G(t)$ and $\phi(t)$ are zero for $t > \beta$ and $\phi(t)$ is zero for $t < \alpha$. We conclude therefore that

$$A_0 = \frac{\int_0^\beta e^{-r_0 t} G(t)\,dt}{\int_\alpha^\beta x\,e^{-r_0 x}\,{}_x p_0 \lambda(x)\,dx}. \tag{3.6.9}$$

Numerical methods of integration are usually necessary to evaluate A_0 from this formula.

More generally, we can show that

$$A_j = \frac{\int_0^\beta e^{-r_j t} G(t) \, dt}{\int_\alpha^\beta x \, e^{-r_j x} {}_x p_0 \lambda(x) \, dx}. \qquad (3.6.10)$$

Further, equation (3.6.1) may be derived from equation (3.6.7) by noting that the Laplace transform of $A_j e^{r_j t}$ is $A_j/(r-r_j)$.

3.7 *Curve fitting the net maternity function*

The method of A. J. Lotka for calculating r_0 was mentioned in section 3.4, and it was shown there that this procedure is equivalent to fitting a normal curve to the net maternity function. Let us proceed further with the normal curve, and assume that

$${}_x p_0 \lambda(x) = \frac{K}{\sigma\sqrt{(2\pi)}} \exp\left\{-\frac{1}{2}\left(\frac{x-\mu}{\sigma}\right)^2\right\}. \qquad (3.7.1)$$

The zeroth, first and second moments are equated, and we discover that

$$\left.\begin{aligned} K &= R_0; \\ \mu &= R_1/R_0; \\ \sigma^2 &= R_2/R_0 - (R_1/R_0)^2. \end{aligned}\right\} \qquad (3.7.2)$$

The mean μ is equal to the cumulant κ_1, and the variance σ^2 is equal to κ_2. We know from section 3.4 therefore that

$$r_0 = \frac{1}{\sigma^2}\{\mu - (\mu^2 - 2\sigma^2 \log R_0)^{\frac{1}{2}}\}. \qquad (3.7.3)$$

The complex roots may be determined in a similar manner. Instead of solving equation (3.4.5), we consider

$$\tfrac{1}{2}r^2\sigma^2 - r\mu + \log R_0 - 2n\pi i = 0, \qquad (3.7.4)$$

where $n = 0, \pm 1, \pm 2, \ldots$. (The real root r_0 is obtained when $n = 0$.) It is soon apparent that the complex root r_n is given by the following formula:

$$r_n = \frac{1}{\sigma^2}[\mu - \{\mu^2 - 2\sigma^2 (\log R_0 - 2n\pi i)\}^{\frac{1}{2}}]. \qquad (3.7.5)$$

Again, it is the negative square root which is relevant to our problem. To see that this is so, consider $\{\mu^2 - 2\sigma^2(\log R_0 - 2n\pi i)\}$. This expression may be written in the form $X(\cos\theta + \mathrm{i}\sin\theta)$, where X is positive[4] and $-\frac{1}{2}\pi < \theta < \frac{1}{2}\pi$. Then the square root is

$$\pm X^{\frac{1}{2}}\{\cos(\tfrac{1}{2}\theta) + \mathrm{i}\sin(\tfrac{1}{2}\theta)\}.$$

Consider the positive sign:

$$\text{Real}\left[\frac{\mu + X^{\frac{1}{2}}\{\cos(\frac{1}{2}\theta) + \mathrm{i}\sin(\frac{1}{2}\theta)\}}{\sigma^2}\right] > \frac{\mu}{\sigma^2} > r_0.$$

But, according to theorem 3.2.1, $r_0 > \text{Real}\,(r_n)$. We conclude therefore that the positive square root is not relevant for our problem. It is easy to show that the negative square root is appropriate.

Other curves have been fitted to the net maternity function. S. D. Wicksell (1931) suggested using a Pearson Type III curve of the form

$$_xp_0\lambda(x) = R_0 \frac{c^k x^{k-1} \mathrm{e}^{-cx}}{\Gamma(k)}. \tag{3.7.6}$$

The parameters c and k are determined by equating first- and second-order moments. The factor R_0 in the right-hand side of formula (3.7.6) ensures that the zeroth moments are correct. Wicksell suggested this curve because of its positive skewness.

Another curve, due to H. Hadwiger (1940), is also positively skewed:

$$_xp_0\lambda(x) = \frac{a}{\sqrt{(\pi x^3)}} \exp\left(ac - bx - \frac{a^2}{x}\right). \tag{3.7.7}$$

The parameters a, b, and c are found by equating the zeroth, first and second moments.

It should be noted that although the Wicksell and Hadwiger curves are positively skewed, the amount of skewness is fixed when the zeroth-, first- and second-order moments are equated. The skewness itself cannot be adjusted.

Explicit formulae for the real root r_0 and the complex roots $\{r_n\}$ may be derived for both these theoretical curves. The curves may be fitted to data using the method of moments outlined above, and approximate values for the roots can then be calculated. Formula (3.5.2) indicates that any one of these distributions will yield an accurate approximation to r_0. But they cannot be relied upon to provide us with accurate values for the complex roots which depend upon the higher-order moments of the net maternity function to a much greater extent. This is perhaps intuitively obvious if one attempts to substitute $\log R_0 + 2n\pi\mathrm{i}$ for $\log R_0$ in formula (3.4.10). Convergence is slower (if at all).

[4] See footnote 3 on page 28.

3.8 *The momentum of population growth*

Hesitation in making contraception available in certain countries is sometimes rationalized by the view that the country should support a much larger population. Fears that total numbers will taper off prematurely in presently high-fertility countries are however quite unfounded, as the following theory will show.

Consider the female component of a stable population with an instantaneous growth rate r_0. We know from section 3.3 that the number of females aged x in the stable population is

$$F(x, 0)\,dx = K\,{}_xp_0\,e^{-r_0 x}\,dx, \tag{3.8.1}$$

and the total female population is obtained by integrating this expression between 0 and ∞. Expression (3.8.1) may be substituted into formula (3.2.3) to obtain

$$G(t) = Ke^{r_0 t}\int_0^\infty e^{-r_0(x+t)}\,{}_{x+t}p_0\,\lambda(x+t)\,dx. \tag{3.8.2}$$

Let us now imagine that a miracle occurs and the age-specific birth-rate function $\lambda(x)$ is suddenly reduced by the factor R_0 so that the population will become stationary. Because of its youthfulness the population will continue to increase for some time, however, and we want to know the ultimate size of the stationary population.

The intrinsic stable growth rate after the change will be zero, and we make use of equation (3.6.9) with r_0 set equal to zero and $\lambda(x)$ replaced by $\lambda(x)/R_0$. Then

$$A_0 = \left(\int_0^\infty G(t)\,dt\right)\Big/\left(\int_0^\infty x\,{}_xp_0\frac{\lambda(x)}{R_0}\,dx\right), \tag{3.8.3}$$

where $G(t)$ is defined by equation (3.8.2).

The average age of child-bearing in the population *before and after* the change in birth rates is

$$\kappa_1 = \int_0^\infty x\,{}_xp_0\frac{\lambda(x)}{R_0}\,dx, \tag{3.8.4}$$

and the birth rate *prior* to the change is

$$b = \left(\int_0^\infty e^{-r_0 x}\,{}_xp_0\lambda(x)\,dx\right)\Big/\left(\int_0^\infty e^{-r_0 x}\,{}_xp_0\,dx\right). \tag{3.8.5}$$

Furthermore, the complete expectation of life at age 0 is

$$\mathring{e}_0 = \int_0^\infty {}_xp_0\,dx. \tag{3.8.6}$$

It is easy to deduce from these results that the size of the stationary population is equal to the number in the population immediately prior to the change multiplied by

$$\frac{b\mathring{e}_0}{r_0\kappa_1}\left(\frac{R_0-1}{R_0}\right). \tag{3.8.7}$$

This elegant formula is due to N. Keyfitz (1971).

Natural populations are not stable populations, but the formula is still surprisingly accurate, as the figures in table 3.8.1 show. In passing, it may be noted that countries of highest present growth have the strongest tendency to grow further. The United States increases only by about a third, and European countries generally by about a fifth. Underdeveloped countries, on the other hand, increase by up to two thirds.

TABLE 3.8.1. *The momenta of certain populations**

	Number in thousands		Per cent increase to ultimate	
Population	Current	Ultimate (numerical calculation)	By numerical calculation	By formula (3.8.7)
Chile 1965	8,584	12,916	50	49
Colombia 1965	17,993	29,786	66	59
Ecuador 1965	5,109	8,518	67	69
Italy 1966	53,128	62,189	17	13
Peru 1963	14,713	23,080	57	53
U.S.A. 1966	195,857	259,490	32	25

* N. Keyfitz (1971).

3.9 *An example*

For a certain population l_x is linear and the ultimate age is 108. The gross maternity function $\lambda(x)$ has the Gaussian form

$$\frac{2}{9\sqrt{(2\pi)}}\exp\left\{-\frac{1}{2}\left(\frac{x-27}{6}\right)^2\right\} \quad (0 \leqslant x < \infty).$$

Evaluate the unique real root r_0.

Solution. We are told that $l_{108} = 0$. Let us assume that $l_0 = 108$. Then $l_x = 108 - x$, and ${}_xp_0 = 1 - x/108$. The net maternity function is

$$\left(1-\frac{x}{108}\right)\frac{(4/3)}{6\sqrt{(2\pi)}}\exp\left\{-\frac{1}{2}\left(\frac{x-27}{6}\right)^2\right\} \quad (0 \leqslant x \leqslant 108),$$

and we require r such that

$$\int_0^{108}\left(1-\frac{x}{108}\right)\frac{(4/3)}{6\sqrt{(2\pi)}}\exp\left\{-\frac{1}{2}\left(\frac{x-27}{6}\right)^2-rx\right\}dx = 1.$$

But r will lie in the neighbourhood of the origin, and an approximate solution may be obtained by replacing the limits of integration by $-\infty$ and $+\infty$. We find that we need to solve the equation

$$(\tfrac{3}{4}+\tfrac{1}{3}r)\exp(27r-18r^2) = \tfrac{3}{4},$$

and it is soon apparent that $r = 0$ is a root. No negative roots exist for this last equation, but at least one large positive root does exist. Clearly $r_0 \doteqdot 0$ is the relevant root for our problem.

3.10 *Exercises*

1 Equation (3.4.5) has two real roots. Prove that the smaller root is the one relevant in the population mathematics context.

2 Prove that equation (3.4.10) is a Taylor series expansion, and give the Lagrangian form of the remainder. *Hint.* Equation (3.4.4) may be written in the form $z(r) = z$, and the function $z(r)$ is monotonic. Use the fact that $dr/dz = 1/(dz/dr)$.

3 A reasonable definition of the generation length T of a population is the time after which the birth rate $B(t)$ has become R_0 times its initial value in the stable, geometrically-increasing population.
Prove that

(i) $T = (\log R_0)/r_0$; and deduce that

(ii) $$T = \kappa_1\left[1-\frac{1}{2}\left(\frac{\kappa_2}{\kappa_1}\right)\left(\frac{\log R_0}{\kappa_1}\right)-\left\{\frac{1}{4}\left(\frac{\kappa_2}{\kappa_1}\right)^2-\frac{1}{6}\left(\frac{\kappa_3}{\kappa_1}\right)\right\}\left(\frac{\log R_0}{\kappa_1}\right)^2\right.$$
$$\left.-\left\{\frac{1}{4}\left(\frac{\kappa_2}{\kappa_1}\right)^3-\frac{1}{4}\left(\frac{\kappa_2\kappa_3}{\kappa_1^{\,2}}\right)-\frac{1}{24}\left(\frac{\kappa_4}{\kappa_1}\right)\right\}\left(\frac{\log R_0}{\kappa_1}\right)^3+o\left(\frac{\log R_0}{\kappa_1}\right)^4\right].$$

4 Use the method of section 3.5 to derive theoretical formulae for the effects of the cumulants of the net maternity function on the generation length T defined in question 3.

5 For a certain population,

$$_xp_0\lambda(x) = e^{-kx}\left(\frac{x^5}{60}+\frac{x^3}{2}\right) \quad (0 \leqslant x < \infty).$$

Prove that the real root r_0 for this population is $-k+\sqrt{2}$.
This distribution was suggested by J. S. Lew (N. Keyfitz, 1968).

6 For a certain population, the force of mortality μ_x is constant for all ages and equal to μ. The maternity function $\lambda(x)$ is also constant for all ages and equal to λ. Determine r_0. Do any complex roots exist?

7 For a certain population, all the roots $\{r_j\}$ are distinct except r_1 which is repeated exactly once. Use the Laplace transform method of section 3.6 to derive formulae suitable for evaluating the constants $\{A_j\}$.

8 Evaluate the Laplace transform of the function

$$B(t) = (A_1 + B_1 t)\, e^{r_1 t}.$$

Use this result to deduce the birth function $B(t)$ for the population described in question 7.

9 The curve suggested by S. D. Wicksell to fit the net maternity function is given in equation (3.7.6). Use the method of moments to determine the constants c and k.

10 Determine the roots $\{r_j\}$ ($j = 0, 1, 2, \ldots$) for a population with a net maternity function of the Wicksell form.

11 Prove that the Wicksell roots $\{r_j\}$ all lie on a circle in the complex plane, and describe the circle.

12 Derive formulae for the quantities $\{A_j\}$ in a population with a Wicksell net maternity function assuming that the population is composed of N females all aged zero at time $t = 0$.
Hint. At time $t = 0$, the age distribution is not continuous. The solution of this question will be facilitated if $F(x, t)\, dx$ in the definition of $G(t)$ is replaced by the Stieltjes form $dF(x, t)$.

13 What is the probability that a newly-born girl-child in a Wicksell-type population will have a grand-daughter via the female line exactly y years later? Generalize this result.

14 In a certain population, the force of mortality μ_x is constant for all ages and equal to μ. The maternity function $\lambda(x)$ is also constant for all ages and equal to λ. Show that this population is a special case of the Wicksell-type population. Deduce a formula for R_0 in terms of μ and λ.

15 Derive the Keyfitz formula (3.8.7).

4
The deterministic theory of H. Bernardelli, P. H. Leslie and E. G. Lewis

4.1 *Introduction*

Deterministic models of population growth exist in two forms: those using a continuous time-variable and a continuous age-scale (following Sharpe and Lotka, 1911), and those using a discrete time-variable and a discrete age-scale. Both types have their advantages, but the discrete formulation is the closer to actuarial practice and is preferable when the age-specific birth and death rates are to be given on the basis of empirical data, rather than as analytical formulae. In this chapter, we discuss the discrete formulation of P. H. Leslie (1945). This is the best-known analysis, and indeed the most detailed. However, two other authors produced the same model earlier. H. Bernardelli gave his in 1941 in the rather obscure *Journal of the Burma Research Society*, and E. G. Lewis published his model in 1942 in *Sankhyā*.

Difficulties arise in the discrete formulation due to the effects of grouping: it is necessary to decide the method of calculating the group survivorship proportions and the group fertility rates; clearly these depend upon the age distribution *within* the age group. Usually, it is assumed that this age distribution follows the stable[1] age distribution. Leslie assumed that the age distribution followed the stationary[2] age distribution which only depends on the life table function l_x.

Small age groups seem preferable, but in practice the decision about grouping is determined to some extent by available data. A time unit of one year for human populations has much to recommend it, and with modern digital computers the computations are very simple. It is surprising how little deterministic results differ when different groupings are used. N. Keyfitz (1964), for example, obtains 1.796 % per annum rate of natural increase for Australian human

[1] The stable age distribution for the continuous-time theory is defined in section 3.3.

[2] The stationary age distribution for the continuous-time theory is obtained from equation (3.3.2) by setting $r_0 = 0$. A stationary population is defined in section 2.10.

females using 1960 data, and a time unit of five years. J. H. Pollard (1966) gives 1.785 % per annum for this rate using a time unit of one year. This latter result was obtained using actual unsmoothed birth rates in 1960 and the 1954 life table. Using the integral equation theory, Keyfitz finds a rate of natural increase of 1.792 % per annum. We shall assume that the time units are small enough to ensure that the effects of grouping are small compared with those due to the other limitations of the model.

4.2 *The matrix method of P. H. Leslie*

The female population only is considered, at discrete points of time $t = 0, 1, 2, \ldots$, and it is broken up into age groups corresponding to the unit intervals of time. Let us assume that there are $m+1$ age groups $0-, 1-, 2-, \ldots, m-$. We then define $n_{x,t}$ to be the number of females in age group x at time t. The proportion of females in age group x at time t surviving to be in age group $x+1$ at time $t+1$ is P_x, which for $x < m$ is strictly positive; $P_m = 0$. Further, F_x denotes the average number of daughters born per female to females in age group x at time t, these daughters surviving to be in age group $0-$ at time $t+1$.

Changes in the male population structure are assumed to be consistent with the assumption of constant fertility rates $\{F_x\}$. Using these definitions, the following relationships are obvious:

$$\left.\begin{aligned} n_{0,\,t+1} &= \sum_{x=0}^{m} F_x n_{x,\,t}, \\ n_{1,\,t+1} &= P_0 n_{0,\,t}, \\ n_{2,\,t+1} &= P_1 n_{1,\,t}, \\ &\vdots \\ n_{m,\,t+1} &= P_{m-1} n_{m-1,\,t}. \end{aligned}\right\} \tag{4.2.1}$$

These equations may be written more neatly in matrix notation:

$$\begin{pmatrix} n_{0,t+1} \\ n_{1,t+1} \\ n_{2,t+1} \\ n_{3,t+1} \\ \vdots \\ n_{m,t+1} \end{pmatrix} = \begin{pmatrix} F_0 & F_1 & F_2 & \cdots & F_{m-1} & F_m \\ P_0 & & & & & \\ & P_1 & & & & \\ & & P_2 & & & \\ & & & \ddots & & \\ & & & & P_{m-1} & 0 \end{pmatrix} \begin{pmatrix} n_{0,t} \\ n_{1,t} \\ n_{2,t} \\ n_{3,t} \\ \vdots \\ n_{m,t} \end{pmatrix}. \tag{4.2.2}$$

That is $\quad \mathbf{n}_{t+1} = \mathbf{M}\mathbf{n}_t. \quad$ (4.2.3)

Hence $\quad \mathbf{n}_t = \mathbf{M}^t\mathbf{n}_0. \quad$ (4.2.4)

Generally $F_j = 0$ for $j > k$, say, and $F_k \neq 0$, and it is useful to study a matrix **A** which is a principal submatrix of **M**. The matrix **A** to be studied is

$$\mathbf{A} = \begin{pmatrix} F_0 & F_1 & F_2 & \cdots & F_{k-1} & F_k \\ P_0 & & & & & \\ & P_1 & & & & \\ & & P_2 & & & \\ & & & \ddots & & \\ & & & & P_{k-1} & 0 \end{pmatrix}, \tag{4.2.5}$$

where $F_k \neq 0$ and $F_{k+j} = 0$ for $j > 0$. This does not reduce the value of the method of investigation, because the individuals alive in the post-reproductive age groups cannot affect the numbers in the reproductive age groups at a later point of time. Equation (4.2.4) becomes

$$\mathbf{n}_t = \mathbf{A}^t \mathbf{n}_0, \tag{4.2.6}$$

where $\mathbf{n}_t$ is now a column vector of dimension $k+1$.

Certain asymptotic results were proved for the Sharpe and Lotka model in section 3.3. Similar results can be proved for the Leslie model, but it is first necessary to discuss the latent roots and latent vectors of the recurrence matrix **A**.

4.3 *The theorem of Perron and Frobenius**

The following theorem, due to O. Perron and G. Frobenius, comes from the theory of matrices.

Theorem 4.3.1. Let **T** be a square matrix with non-negative elements only, and such that all the elements of $\mathbf{T}^N$ are positive for some positive integer N. Then **T** has a positive latent root of algebraic multiplicity one, which corresponds to a latent column vector **u** and to a latent row vector **v**′, both of which have only positive elements. This latent root is greater in absolute size than any other latent root of **T**.

The Leslie matrix **A** has a very simple form, and it is not necessary to use this general theorem in the present analysis. The theorem may be required, however, to deal with certain other population models, and we shall therefore demonstrate its use with the Leslie model.

Most species reproduce over a period of their life, and so two consecutive F_j are usually non-zero. This is sufficient to ensure that all the elements of $\mathbf{A}^N$ are positive for some positive integer N and the process is said to be positive regular (T. E. Harris, 1963, 37–8). The

Perron–Frobenius theorem can then be applied. An artificial example of a non-regular process is given in exercise 1 of section 4.13.

The condition that two consecutive F_j be positive is sufficient but not necessary for positive regularity. A necessary and sufficient condition for the Leslie matrix $\mathbf{A}$ to be positive regular is provided by theorem 4.3.4. This result depends upon the following two theorems from number theory.

Theorem 4.3.2. If D is the greatest common divisor of n positive integers $\{A_i\}$ $(i = 1, \ldots, n)$, then there exist integers $\{a_i^*\}$ $(i = 1, \ldots, n)$ such that

$$\sum_1^n a_i^* A_i = D.$$

Proof. We consider linear combinations of the $\{A_i\}$ of the form $\Sigma A_i a_i$, where the $\{a_i\}$ range over all the integers (positive, zero and negative). Clearly the set of integers $\{\Sigma A_i a_i\}$ includes positive, zero and negative values. We select the integers $\{a_i^*\}$ such that $L = \Sigma A_i a_i^*$ is the least positive integer in the set $\{\Sigma A_i a_i\}$.

Let us assume that L is *not* a divisor of A_1. Then integers Q and R exist such that $A_1 = LQ + R$, and $0 < R < L$. We can see that

$$\begin{aligned} R &= A_1 - LQ \\ &= A_1 - Q\Sigma A_i a_i^* \\ &= A_1(1 - Qa_1^*) - \sum_{i \neq 1} A_i(Qa_i^*) \end{aligned}$$

which is included in the set $\{\Sigma A_i a_i\}$. But R is positive and less than L, and this contradicts the definition of L. We conclude that L is a divisor of A_1, and more generally L is a divisor of A_i $(i = 1, \ldots, n)$.

D is the greatest common divisor of the $\{A_i\}$. We can therefore write $A_i = DX_i$. Then

$$L = \Sigma A_i a_i^* = D\Sigma X_i a_i^*.$$

That is, D divides L, and therefore $D \leqslant L$. It is impossible for D to be less than L because D is the greatest common divisor of the $\{A_i\}$, and it follows therefore that

$$D = L = \Sigma A_i a_i^*.$$

This completes the proof of the theorem.

Corollary 4.3.2. If the greatest common divisor of n positive integers $\{A_i\}$ $(i = 1, \ldots, n)$ is one, then integers $\{a_i^*\}$ exist such that $\Sigma A_i a_i^* = 1$.

Theorem 4.3.3. If the greatest common divisor of n positive integers A_i $(i = 1, \ldots, n)$ is one, then there exists a positive integer Q_0 such that all integers greater than Q_0 may be written in the form $\Sigma A_i a_i$ where the $\{a_i\}$ are non-negative integers.

Proof. According to corollary 4.3.2, we can select integers $\{a_i^*\}$ such that $\Sigma A_i a_i^* = 1$. It is clear that some of the $\{a_i^*\}$ are negative. Let us imagine that a_1^* and a_2^* are negative and all the others are non-negative. Consider an integer

$$Q \geqslant -a_1^* A_1(-a_1^* A_1 - 1) - a_2^* A_2(-a_2^* A_2 - 1).$$

We may write

$$Q = X_1(-a_1^* A_1) + X_2(-a_2^* A_2) + Y,$$

where $X_1 \geqslant -a_1^* A_1 - 1,$

$X_2 \geqslant -a_2^* A_2 - 1,$

and $0 \leqslant Y < \min(-a_1^* A_1, -a_2^* A_2).$

Then

$$\begin{aligned} Q &= X_1(-a_1^* A_1) + X_2(-a_2^* A_2) + Y \\ &= X_1(-a_1^* A_1) + X_2(-a_2^* A_2) + Y(\Sigma a_i^* A_i) \\ &= (X_1 - Y)(-a_1^* A_1) + (X_2 - Y)(-a_2^* A_2) + \sum_{i>2} (Y a_i^*) A_i \\ &= \Sigma a_i A_i, \end{aligned}$$

where the $\{a_i\}$ are non-negative integers. The generalization of the proof if obvious.

Corollary 4.3.3. If the greatest common divisor of n positive integers $\{A_i\}$ $(i = 1, \ldots, n)$ is D, then there exists a positive integer Q_0 such that all multiples of D of the form QD greater than $Q_0 D$ may be written in the form $QD = \Sigma A_i a_i$, where the $\{a_i\}$ are non-negative integers. Integers which are not multiples of D cannot be written in this form.

Theorem 4.3.4. The Leslie matrix **A** is positive regular if and only if the greatest common divisor of the numbers of the columns containing positive fertility measures is one.

Proof. Let us assume that $F_{r_1}, F_{r_2}, \ldots, F_{r_n}$ are positive $(r_n \equiv k)$, and that the greatest common divisor of $r_1 + 1, \ldots, r_n + 1$ is one.

If we refer to age i as state i and we also borrow the word 'accessible' from Markov chain theory, it is clear that state r_1 is accessible from state 0 in the product matrix $\mathbf{A}^{r_1}$. It follows that state 0 is accessible from itself in the product matrix $\mathbf{A}^{r_1+1}$. More generally, state 0 is accessible from itself in the product matrix $\mathbf{A}^t$ whenever $t = \Sigma a_i(r_i+1)$, and the $\{a_i\}$ are non-negative integers.

Now the greatest common divisor of the $\{r_i+1\}$ is one. It follows therefore from theorem 4.3.3 that an integer T_0 exists such that for all $t > T_0$, state 0 is accessible from itself in the product matrix $\mathbf{A}^t$. That is, the elements of the first row of the matrix $\mathbf{A}^{t+k}$ will be strictly positive for all $t > T_0$. (The number k is defined in section 4.2.) The form of the Leslie matrix with strictly positive elements $\{P_j\}$ down the subdiagonal will ensure that all the elements of $\mathbf{A}^{t+2k}$ will be positive whenever $t > T_0$. That is, the Leslie matrix is positive regular. This proves the sufficiency of the condition. Corollary 4.3.3 gives us the necessity of the condition.

Two consecutive integers must be relatively prime, and this gives us the result quoted earlier in this section, which can also be proved from first principles.

4.4 *The latent roots of the Leslie matrix A*

It is possible to prove that the matrix **A** has a positive dominant latent root λ_0 of algebraic multiplicity one without using the theorem of Perron and Frobenius when at least two consecutive F_j are positive. Most species reproduce over a period of their life, so two consecutive F_j will usually be non-zero. We first note that the characteristic equation of the matrix **A** is

$$\lambda^{k+1} - F_0\lambda^k - P_0F_1\lambda^{k-1} - P_0P_1F_2\lambda^{k-2} - \ldots - P_0P_1\ldots P_{k-1}F_k = 0. \quad (4.4.1)$$

A has no zero latent roots. For $\lambda \neq 0$, this equation may be written

$$F_0\lambda^{-1} + P_0F_1\lambda^{-2} + P_0P_1F_2\lambda^{-3} + \ldots + P_0P_1\ldots P_{k-1}F_k\lambda^{-(k+1)} = 1. \quad (4.4.2)$$

That is, $f(\lambda) = 1$. (4.4.3)

It is clear that $f(\lambda)$ decreases strictly monotonically from ∞ to 0 as λ increases from 0+ to ∞. Hence there is only one real positive root of multiplicity one.

Let λ_j be any other latent root, and write $\lambda_j^{-1} = e^{\alpha+i\beta}$, where α and β are real and positive, and $\beta \neq 2r\pi$. That is, λ_j must be negative or complex. We then have

$$\lambda_j^{-n} = \{e^{\alpha}(\cos\beta + i\sin\beta)\}^n = e^{n\alpha}\{\cos(n\beta) + i\sin(n\beta)\}.$$

Substituting for λ_j^{-n} in equation (4.4.2) and equating the real parts of the left-hand side and right-hand side, we obtain

$$F_0 e^{\alpha} \cos\beta + P_0 F_1 e^{2\alpha} \cos(2\beta) + \ldots + P_0 P_1 \ldots P_{k-1} F_k e^{(k+1)\alpha} \cos\{(k+1)\beta\} = 1. \quad (4.4.4)$$

Since $\beta \neq 2r\pi$, $\cos(n\beta)$ and $\cos\{(n+1)\beta\}$ cannot both be unity, and so either or both must be less than unity. A comparison of equation (4.4.4) with equation (4.4.2) shows that e^{α} must be greater than λ_0^{-1}, and we conclude that

$$|\lambda_j| < \lambda_0. \quad (4.4.5)$$

The above proof should be compared with the one in section 3.2.

It is very simple to determine the right and left latent vectors of **A** corresponding to the dominant latent root λ_0. Let these vectors be **x** and **y**′ respectively. Consider **x** first.

$$\mathbf{Ax} = \lambda_0 \mathbf{x}. \quad (4.4.6)$$

We may arbitrarily select x_0 equal to one. Then, from equation (4.4.6),

$$P_0 x_0 = \lambda_0 x_1,$$

so that $x_1 = P_0/\lambda_0$,

and by induction

$$x_j = P_0 P_1 \ldots P_{j-1} \lambda_0^{-j}. \quad (4.4.7)$$

Clearly all the elements of **x** are positive. Formula (4.4.7) should be compared with equation (3.3.2) in the continuous-time theory. The integral in the denominator of equation (3.3.2) is a scaling factor and should be ignored in this comparison.

It is possible to proceed in a similar manner with **y**′, setting y_0 equal to one. We find that

$$y_j = \sum_{n=j}^{k} (P_j P_{j+1} \ldots P_{n-1}) F_n \lambda_0^{j-n-1}. \quad (4.4.8)$$

The form of the elements of **y** should be noted. It is soon apparent that the element y_j is equal to the discounted future births to a female aged j; the discount rate corresponds to a rate of interest $\lambda_0 - 1$ (i.e. the intrinsic rate of increase per annum of the population). This result seems to have been first noted by R. A. Fisher (1930), who called y_j *the reproductive value* of a woman in age group j. Formula (4.4.8) should be compared with the continuous-time reproductive value at age u:

$$\int_u^{\beta} {}_{t-u}p_u \lambda(t)\, e^{r_0(u-t)}\, dt. \quad (4.4.9)$$

It was noted above that all the elements of $\mathbf{x}$ are positive, and it is easy to see that this is also true of the elements of $\mathbf{y}$. It is possible therefore to scale $\mathbf{x}$ and $\mathbf{y}$ so that

$$\mathbf{y}'\mathbf{x} = 1. \tag{4.4.10}$$

This equation will be satisfied if we redefine $\mathbf{x}$ as follows:

$$x_j = \frac{P_0 P_1 \dots P_{j-1} \lambda_0^{-j}}{\sum_{n=0}^{k} (n+1)(P_0 P_1 \dots P_{n-1}) F_n \lambda_0^{-(n+1)}}. \tag{4.4.11}$$

It will be assumed hereafter that vectors $\mathbf{x}$ and $\mathbf{y}'$ are those defined in equations (4.4.11) and (4.4.8) respectively.

4.5 *The asymptotic stable age distribution*

The results of sections 4.3 and 4.4 lead us to a stable age distribution because the dominant latent root λ_0 of $\mathbf{A}$ will ultimately swamp the effects of the other latent roots. To prove that this is true, we require the following lemma.

Lemma 4.5.1. Let $\mathbf{T}$ be a square matrix of order $k+1$ with an algebraically simple strictly dominant latent root ρ_0. Let $\mathbf{u}$ and $\mathbf{v}'$ be the corresponding right and left latent vectors. It is possible to choose $\mathbf{u}$ and $\mathbf{v}'$ such that $\mathbf{v}'\mathbf{u} = 1$, and then for large values of n,

$$\mathbf{T}^n/\rho_0^n = \mathbf{u}\mathbf{v}' + \mathbf{E}(n),$$

where the moduli of the elements of $\mathbf{E}(n)$ are no greater than a number of order $n^{k-1}|\rho_1/\rho_0|^n$ and $|\rho_1| < \rho_0$.

Proof. Let ρ_1 be a latent root with the second largest modulus. There exists a non-singular matrix $\mathbf{H}$ such that

$$\mathbf{T} = \mathbf{H}\begin{pmatrix} \rho_0 & \mathbf{0}' \\ \mathbf{0} & \mathbf{Z} \end{pmatrix}\mathbf{H}^{-1},$$

where $\mathbf{Z}$ consists of diagonally arranged blocks like

$$\mathbf{Z}(r) = \begin{pmatrix} \rho_r & 1 & & \\ & \ddots & \ddots & \\ & & \ddots & 1 \\ & & & \rho_r \end{pmatrix},$$

and each ρ_r is a latent root of $\mathbf{T}$ (Turnbull and Aitken, 1951).

Let us consider a 3×3 diagonal block as an example.

If $$\mathbf{Z}(1) = \begin{pmatrix} \rho_1 & 1 & 0 \\ 0 & \rho_1 & 1 \\ 0 & 0 & \rho_1 \end{pmatrix},$$

then $$\{\mathbf{Z}(1)\}^5 = \begin{pmatrix} \rho_1^5 & 5\rho_1^4 & 10\rho_1^3 \\ 0 & \rho_1^5 & 5\rho_1^4 \\ 0 & 0 & \rho_1^5 \end{pmatrix}.$$

The upper triangular elements of $\{\mathbf{Z}(1)\}^5$ belong to the binomial expansion of $(\rho_1+1)^5$. In the more general situation, $\mathbf{Z}(1)$ is a $p \times p$ matrix, and we see that for large n, the modulus of the largest element of $\{\mathbf{Z}(1)\}^n$ will be of order $n^{p-1}|\rho_1|^n$.

The matrix $\mathbf{T}$ is of dimension $(k+1) \times (k+1)$, and the worst possible case occurs when $\mathbf{Z}(1)$ is of dimension $k \times k$. It follows that

$$\mathbf{T}^n/\rho_0^n = \mathbf{H}\begin{pmatrix} 1 & \mathbf{0}' \\ \mathbf{0} & \mathbf{0} \end{pmatrix}\mathbf{H}^{-1} + \mathbf{E}(n), \tag{4.5.1}$$

where the moduli of the elements of $\mathbf{E}(n)$ are all less than or equal to a number of order $n^{k-1}|\rho_1/\rho_0|^n$.

The next step is to note that

$$\lim_{n\to\infty} \mathbf{T}^n/\rho_0^n = \mathbf{H}\begin{pmatrix} 1 & \mathbf{0}' \\ \mathbf{0} & \mathbf{0} \end{pmatrix}\mathbf{H}^{-1},$$

and because $\mathbf{u}$ is the right eigenvector of $\mathbf{T}$ corresponding to ρ_0, we must have

$$\mathbf{H}^{-1}\mathbf{u} = \begin{pmatrix} 1 & \mathbf{0} \\ \mathbf{0} & \mathbf{0} \end{pmatrix}\mathbf{H}^{-1}\mathbf{u} = \alpha\begin{pmatrix} 1 \\ \mathbf{0} \end{pmatrix}.$$

In a similar manner $\mathbf{v}'\mathbf{H} = \beta(1, \mathbf{0}')$.

The constants α and β are both non-zero, and it is therefore possible to normalize $\mathbf{u}$ and $\mathbf{v}'$ so that

$$\mathbf{v}'\mathbf{u} = \mathbf{v}'\mathbf{H}\mathbf{H}^{-1}\mathbf{u} = \alpha\beta = 1.$$

We now note that

$$\mathbf{u}\mathbf{v}' = \mathbf{H}\begin{pmatrix} 1 & \mathbf{0} \\ \mathbf{0} & \mathbf{0} \end{pmatrix}\mathbf{H}^{-1},$$

and it follows from equation (4.5.1) that

$$T^n/\rho_0^n = \mathbf{u}\mathbf{v}' + \mathbf{E}(n).$$

The proof of the lemma is now complete.

Applying lemma 4.5.1 to the Leslie population,

$$\mathbf{n}_t = \mathbf{A}^t\mathbf{n}_0 = \lambda_0^t(\mathbf{y}'\mathbf{n}_0)\,\mathbf{x} + o(\lambda_0^t), \tag{4.5.2}$$

and the population has an asymptotic stable age distribution. The vectors $\mathbf{x}$ and $\mathbf{y}'$ are the right and left eigenvectors of matrix $\mathbf{A}$ corresponding to the dominant latent root λ_0, and these eigenvectors are defined by equations (4.4.11) and (4.4.8) respectively.

Result (4.5.2) should be compared with result (3.3.1). The rth element of vector $\mathbf{n}_t$ may be found by substituting the formulae for $\mathbf{y}'$ and $\mathbf{x}$ into the right-hand side of equation (4.5.2):

$$n_{r,t} \simeq \left\{ \frac{\sum_{j=0}^{k} \sum_{m=j}^{k} (P_j \dots P_{m-1}) F_m \lambda_0^{j-m-1} n_{j,0}}{\sum_{m=0}^{k} (m+1)(P_0 \dots P_{m-1}) F_m \lambda_0^{-(m+1)}} \right\} (P_0 \dots P_{r-1}) \lambda_0^{t-r}. \tag{4.5.3}$$

Formulae (3.6.9) and (3.2.3) for A_0 and $G(u)$ may be substituted into the right-hand side of equation (3.3.1) to obtain the corresponding formula in the continuous analysis:

$$F(x,t) \simeq \left\{ \frac{\int_{u=0}^{\beta} \int_{T=u}^{\beta} (l_T/l_u)\, \lambda(T)\, e^{r_0(u-T)} F(u,0)\, dT\, du}{\int_{u=0}^{\beta} u\, {}_u p_0\, \lambda(u)\, e^{-r_0 u}\, du} \right\} \times {}_x p_0\, e^{r_0(t-x)}. \tag{4.5.4}$$

The results for both these models should be compared with the Malthus result given in section 3.1.

4.6 *The recurrence equation approach to the discrete-time model*

It is possible to analyse the discrete-time model without making use of matrix methods. Such an approach is instructive in that the procedure adopted is the discrete-time version of the continuous-time analysis of sections 3.2 and 3.3.

We have defined $n_{x,t}$ to be the number of females at time t whose ages lie between x and $x+1$, and $n_{0,t}$ to be the number of females who were born during the time interval $(t-1, t)$ and who survived until time t. From these definitions, and the definition of P_x, it is clear that

$$\left.\begin{aligned} n_{x,t} &= (P_0 P_1 \dots P_{x-1})\, n_{0,t-x} \quad (0 \leqslant x \leqslant t), \\ \text{and} \qquad n_{x,t} &= (P_{x-t} \dots P_{x-1})\, n_{x-t,0} \quad (x > t). \end{aligned}\right\} \tag{4.6.1}$$

Then
$$\begin{aligned} n_{0,t+1} &= \sum_{x=0}^{\infty} n_{x,t} F_x \\ &= \sum_{x=0}^{t} (P_0 P_1 \dots P_{x-1}) F_x n_{0,t-x} + \sum_{x=1}^{\infty} (P_x \dots P_{x+t-1}) F_{x+t} n_{x,0}. \end{aligned}$$

That is
$$n_{0,t+1} = \sum_{x=0}^{t} (P_0 P_1 \dots P_{x-1}) F_x n_{0,t-x} + G_t, \tag{4.6.2}$$
where
$$G_t = \sum_{x=1}^{\infty} n_{x,0}(P_x \dots P_{x+t-1}) F_{x+t}. \tag{4.6.3}$$
This is the basic recurrence equation for the Leslie population model.

Let us continue to follow the analysis of section 3.2. Reproduction takes place over only part of the life span of a female, and we assume that $F_x = 0$ for all $x > k$. Therefore
$$G_t = 0, (t > k). \tag{4.6.4}$$
Consider $t > k$. Equation (4.6.2) becomes a homogeneous linear recurrence relation of order $k+1$. It is standard procedure to make a trial solution of the form
$$n_{0,t} = c\lambda^t. \tag{4.6.5}$$
When this is substituted in equation (4.6.2), we obtain
$$c\lambda^{t+1} = \sum_{x=0}^{k} c(P_0 P_1 \dots P_{x-1}) F_x \lambda^{t-x}.$$
Then λ is the solution of the polynomial equation
$$\lambda^{k+1} - F_0\lambda^k - P_0 F_1 \lambda^{k-1} - \dots - P_0 P_1 \dots P_{k-1} F_k = 0. \tag{4.6.6}$$
This equation is of course the characteristic equation of the Leslie matrix **A**.

Theorem 4.6.1. The polynomial equation (4.6.6) has one real positive root $\lambda = \lambda_0$ of algebraic multiplicity one. The other roots
$$\{\lambda_i\} \quad (i = 1, 2, \dots, k)$$
are either negative or complex, and $\lambda_0 > |\lambda_i|$.

This result was proved (without the use of matrix methods) in section 4.4, making the not unreasonable assumption that at least two consecutive $\{F_j\}$ are non-zero.

From the theory of homogeneous linear recurrence equations, we know that if λ_i is a root of equation (4.6.6) of multiplicity m, the recurrence equation has a solution of the form
$$n_{0,t} = (C_0 + C_1 t + \dots + C_{m-1} t^{m-1}) \lambda_i^t. \tag{4.6.7}$$
According to the results of theorem 4.6.1, such solutions are negligible compared with $\lambda_0{}^t$ for large t, and we conclude that
$$n_{0,t} \cong c_0 \lambda_0{}^t \quad \text{for large } t. \tag{4.6.8}$$

This result may be substituted in equation (4.6.1) to obtain a stable age distribution for the population:

$$n_{x,t} \cong (P_0 P_1 \ldots P_{x-1})\, c_0 \lambda_0^{\,t-x}. \tag{4.6.9}$$

The proportion of females in the age group $x-$ is therefore given by

$$\frac{n_{x,t}}{\sum_{u=0}^{\infty} n_{u,t}} \cong \frac{(P_0 P_1 \ldots P_{x-1})\, \lambda_0^{-x}}{\sum_{u=0}^{\infty} (P_0 \ldots P_{u-1})\, \lambda_0^{-u}}. \tag{4.6.10}$$

The methods used in this section should be compared with those in sections 3.2 and 3.3. Equations (4.6.1) to (4.6.6) correspond with equations (3.2.1) to (3.2.6) respectively, and equations (4.6.7) to (4.6.10) correspond with (3.2.13), (3.2.14), (3.3.1) and (3.3.2) respectively. The proof of theorem 4.6.1 should be compared with the proof of theorem 3.2.1.

4.7 *The sensitivity of the intrinsic growth rate to changes in the age-specific birth and death rates*

Let us now examine the effect on λ_0 of a change in the value of F_j, the other parameters being kept constant. The Leslie characteristic equation (4.4.2) may be written in the form

$$\phi(\lambda_0, F_j) = 1,$$

so that
$$\frac{\partial \phi}{\partial F_j} + \frac{\partial \phi}{\partial \lambda_0}\frac{d\lambda_0}{dF_j} = 0,$$

and
$$\frac{d\lambda_0}{dF_j} = -\frac{\partial \phi}{\partial F_j}\bigg/\frac{\partial \phi}{\partial \lambda_0}. \tag{4.7.1}$$

The partial derivatives are available immediately from equation (4.4.2):

$$\frac{\partial \phi}{\partial F_j} = P_0 P_1 \ldots P_{j-1} \lambda_0^{-(j+1)}. \tag{4.7.2}$$

$$\frac{\partial \phi}{\partial \lambda_0} = -\{F_0 \lambda_0^{-2} + 2P_0 F_1 \lambda_0^{-3} + \ldots + (k+1)(P_0 \ldots P_{k-1}) F_k \lambda_0^{-(k+2)}\}.$$

When the partial derivative with respect to λ_0 is compared with the continuous-time average age of childbearing κ_1 defined in equation (3.8.4), we see that

$$\frac{\partial \phi}{\partial \lambda_0} \doteqdot -R_0 \kappa_1 \lambda_0^{-(\kappa_1+1)}. \tag{4.7.3}$$

so that
$$\frac{d\lambda_0}{dF_j} \doteqdot \{(P_0P_1 \ldots P_{j-1})\,\lambda_0^{\kappa_1-j}\}/(R_0\kappa_1). \tag{4.7.4}$$

A numerical example is instructive. According to table 3.5.1, the average age of childbearing in the United States of America in 1964 was about 27, and the net reproduction rate R_0 was 1.526. The probability of survival from 0 to age 27 should be a little better than that for Australian males, say 0.96. Then $d\lambda_0/dF_{27}$ is approximately 0.024. This figure may seem large, but it must be remembered that the figure corresponds to a change of one in F_{27}. The largest value of F_j for the American population will be about 0.25 in the vicinity of age 23. The following formula gives the proportionate change in λ_0 for a proportionate change in F_j.

$$\frac{\Delta\lambda_0}{\lambda_0}\bigg/\frac{\Delta F_j}{F_j} \doteqdot \{(P_0P_1 \ldots P_{j-1})\,F_j\lambda_0^{\kappa_1-j-1}\}/(R_0\kappa_1). \tag{4.7.5}$$

Let us now examine the effect on λ_0 of a small change in the survival rate P_j. Mathematically, a change of one per cent in P_j is equivalent to a change of the same proportion in each of the fertility rates $F_{j+1}, F_{j+2}, \ldots, F_k$. We can conclude therefore that

$$\frac{\Delta\lambda_0}{\lambda_0}\bigg/\frac{\Delta P_j}{P_j} \doteqdot \{(P_0P_1 \ldots P_j)\,\lambda_0^{\kappa_1-j-1}y_{j+1}\}/(R_0\kappa_1), \tag{4.7.6}$$

where y_{j+1}, the reproductive value of a female aged $j+1$, is defined by equation (4.4.8).

L. Demetrius (1969) and L. A. Goodman (1971) have studied these problems in some detail, and the interested reader should refer to their original papers.

4.8 *Some further investigations by P. H. Leslie*

In the discrete-time analysis of sections 4.1, 4.2, 4.4 and 4.5, the quantities $\{P_i\}$ and $\{F_i\}$ are assumed to be independent of the population size, and it follows that the population increases geometrically once the stable age distribution is reached. Ignoring the age structure, this situation corresponds to the continuous-time differential equation (3.1.1), which was studied by A. J. Lotka in 1907. Leslie was concerned primarily with animal populations, and in his 1948 paper was interested in the effects of overcrowding. He therefore sought an analogue in discrete time for the logistic population defined by equation (3.1.2). He considered two main cases separately:

(*a*) mortality affected by a factor independent of age, fertility remaining constant; and (*b*) fertility affected by a factor independent of age, mortality remaining constant.

The models he used are somewhat unrealistic, and since our interest lies elsewhere, we do not discuss the problem further. Leslie also investigated some aspects of the predator–prey problem.

4.9 *Some deterministic extensions*

Variations of Leslie's methods have often been used to analyse numerical data in specific scientific problems. J. H. Darwin and R. M. Williams (1964), for example, studied the effect of time of hunting on the size of a rabbit population using these techniques; J. Gani, in 1963, produced formulae for projecting enrolments and degrees in universities. Almost invariably, the methods used in such circumstances have been deterministic. Three deterministic population models, which make allowances for immigration will now be described.

Immigration model (1). Let us assume that each year a constant immigration vector $\mathbf{b}$ is added to the population. Then

$$\mathbf{n}_{t+1} = \mathbf{A}\mathbf{n}_t + \mathbf{b}, \tag{4.9.1}$$

and

$$\mathbf{n}_t = \mathbf{A}^t\mathbf{n}_0 + (\mathbf{I}-\mathbf{A})^{-1}(\mathbf{I}-\mathbf{A}^t)\,\mathbf{b}, \tag{4.9.2}$$

assuming that $\mathbf{I}-\mathbf{A}$ is non-singular, which is equivalent to assuming that $\lambda_0 \neq 1$.

The asymptotic behaviour of the population is soon apparent. If, for example, $\lambda_0 < 1$, $\mathbf{A}^t \to \mathbf{0}$ and we find that

$$\mathbf{n}_t \simeq (\mathbf{I}-\mathbf{A})^{-1}\mathbf{b}. \tag{4.9.3}$$

Immigration model (2). Let us assume that, instead, a population vector proportional to the actual population vector is added to the population each year. That is

$$\mathbf{n}_{t+1} = (\mathbf{A}+\alpha\mathbf{I})\,\mathbf{n}_t. \tag{4.9.4}$$

It may be noted that the matrix $(\mathbf{A}+\alpha\mathbf{I})$ has the same latent vectors as $\mathbf{A}$, and has a dominant latent root $(\lambda_0+\alpha)$. The asymptotic behaviour is readily determined.

Immigration model (3). In this model, it will be assumed that immigration may be represented by a combination of the following three factors:

(i) an immigration vector whose elements are proportional to those in the population;

(ii) a constant immigration vector; and

(iii) an immigration vector whose elements grow exponentially at the same rate with time.

Mathematically, the model may be represented by the recurrence equation

$$\mathbf{n}_{t+1} = \mathbf{A}\mathbf{n}_t + \alpha\mathbf{n}_t + \mathbf{b} + \beta^{t+1}\mathbf{c}, \tag{4.9.5}$$

which leads to the asymptotic formula

$$\mathbf{n}_t = (\lambda_0 + \alpha)^t \mathbf{l} + \beta^t \mathbf{m} + \mathbf{p} + o\{(\lambda_0 + \alpha)^t\},$$

where $\mathbf{l}$, $\mathbf{m}$, and $\mathbf{p}$ are certain fixed vectors. The asymptotic values of $\mathbf{n}_t$ are obvious for the various possible cases.

Many other simple models are possible. The problem of immigration will be discussed again in chapter 10, when stochastic fluctuations will be taken into account.

4.10 *Weak ergodicity**

It was proved in section 4.5 that the Leslie population has an asymptotic age distribution which is independent of the initial age structure. This ergodic property was proved under the strong conditions of unchanging age-specific fertility and mortality rates. In the real world, fertility and mortality rates undergo *secular changes* (i.e. changes over time), and A. J. Coale (1957) conjectured that this property of a population tending to forget its initial age structure should be true under somewhat weaker conditions. Such a result was proved by A. Lopez in 1961. He proved that two populations which are arbitrarily different in their age structure will tend to adopt the same age distribution as each other with the passage of time if they are both subjected to the same regimes of fertility and mortality which are *not* assumed to be unchanging.

Lopez considered two female populations with the age distribution vectors $\mathbf{m}_t = (m_{j,t})$ and $\mathbf{n}_t = (n_{j,t})$ respectively at time t. Both populations are subjected to the same sequence of mortality and fertility rates, and we define the Leslie matrix at time t as

$$\mathbf{A}^{(t)} = \begin{pmatrix} F_0^{(t)} & F_1^{(t)} & \cdots & F_{k-1}^{(t)} & F_k^{(t)} \\ P_0^{(t)} & & & & \\ & P_1^{(t)} & & & \\ & & \ddots & & \\ & & & P_{k-1}^{(t)} & 0 \end{pmatrix}. \tag{4.10.1}$$

The time variable t assumes the values 1, 2, 3, etc.

Lopez makes several assumptions:

(i) There exists a positive number ϵ such that for all j and all t, $F_j^{(t)} > \epsilon > 0$, or else $F_j^{(t)} = 0$.

(ii) If for a certain j, $F_j^{(t)}$ is positive for any particular value of t, then $F_j^{(t)}$ is positive for all t.

(iii) The greatest common divisor of the column numbers containing non-zero fertility measures is one. (Column j contains the fertility measure for age $j-1$.)

(iv) The survival rates $\{P_j^{(t)}\}$ $(0 \leqslant j \leqslant k)$ are all greater than the ϵ defined in assumption (i).

(v) All the elements of $\mathbf{m}_0$ and $\mathbf{n}_0$ are strictly positive.

(vi) The fertility measures $\{F_j^{(t)}\}$ have an upper bound.

From assumptions (i) and (iv), it is clear that all the columns in $\mathbf{A}^{(t)}$ have at least one positive entry greater than the positive number ϵ, and because of the form of the Leslie matrix, no column can contain more than two positive entries. We also note that the ratios $\{m_{i,0}/m_{j,0}\}$ and $\{n_{i,0}/n_{j,0}\}$ are bounded above and below by two positive numbers.

We need to prove that

$$\lim_{t\to\infty}\left(\frac{m_{i,t}}{n_{i,t}}-\frac{m_{j,t}}{n_{j,t}}\right) = 0. \qquad (4.10.2)$$

Assumptions (ii) and (iii) allow us to apply theorem 4.3.4 to the product matrix

$$\mathbf{A}^{(t+N)}\mathbf{A}^{(t+N-1)}\ldots\mathbf{A}^{(t+1)},$$

and we deduce that for a sufficiently large N, N_0 say, all the elements of the product matrix are positive. It is convenient to define 'lumped' matrices $\{\mathbf{H}^{(T)}\}$ as follows:

$$\mathbf{H}^{(T)} = \mathbf{A}^{(TN_0)}\mathbf{A}^{(TN_0-1)}\ldots\mathbf{A}^{(TN_0-N_0+1)}. \qquad (4.10.3)$$

These matrices are all strictly positive. Using assumption (vi), it is not difficult to reason that the elements of $\mathbf{H}^{(T)} = (h_{ij}^{(T)})$ have a finite upper bound. These elements also have a positive lower bound, and we see that $h_{ij}^{(T)}$ is greater than ϵ^{N_0}. Thus

$$\epsilon^{N_0} < h_{ij}^{(T)} < M, \text{ say.} \qquad (4.10.4)$$

We now consider the 'lumped' recurrence relation, and for convenience, we shall write $\mathbf{m}^{(T)}$ (with elements $\{m_j^{(T)}\}, j = 0, 1, 2, \ldots, k$) for $\mathbf{m}_{TN_0}$ and $\mathbf{n}^{(T)}$ for $\mathbf{n}_{TN_0}$. Then

$$\left.\begin{aligned} m_i^{(T+1)} &= \sum_{j=0}^{k} h_{ij}^{(T+1)} m_j^{(T)}, \\ \text{and} \qquad n_i^{(T+1)} &= \sum_{j=0}^{k} h_{ij}^{(T+1)} n_j^{(T)}. \end{aligned}\right\} \tag{4.10.5}$$

We select the minimum and maximum ratios of the form $m_i^{(T)}/n_i^{(T)}$ and denote these by r_T and R_T respectively. It is soon apparent that

$$r_T = \frac{m_{L(T)}^{(T)}}{n_{L(T)}^{(T)}} \leqslant \frac{m_i^{(T+1)}}{n_i^{(T+1)}} \leqslant \frac{m_{U(T)}^{(T)}}{n_{U(T)}^{(T)}} = R_T \tag{4.10.6}$$

for all i. The symbol $L(T)$ indicates the number of the elements of $\mathbf{m}^{(T)}$ and $\mathbf{n}^{(T)}$ giving the lowest ratio at time T. The symbol $U(T)$ indicates the number of the elements with the highest (or uppermost) ratio at time T. The inequality is brought about by the fact that the same linear transformation is applied to both populations.

Setting $i = L(T+1)$ in equation (4.10.6) leads to the result

$$r_T \leqslant r_{T+1} \leqslant R_T, \tag{4.10.7}$$

and if we set i equal to $U(T+1)$ we obtain

$$R_T \geqslant R_{T+1} \geqslant r_T. \tag{4.10.8}$$

From these inequalities we deduce that

$$r_0 \leqslant r_1 \leqslant r_2 \leqslant \ldots \leqslant r_n \leqslant \ldots \leqslant R_0 \tag{4.10.9}$$

and

$$R_0 \geqslant R_1 \geqslant R_2 \geqslant \ldots \geqslant R_n \geqslant \ldots \geqslant r_0. \tag{4.10.10}$$

The initial populations were chosen in such a manner that both r_0 and R_0 were positive and finite. Both the sequences $\{r_i\}$ and $\{R_i\}$ therefore converge, and it remains to be proved that they have the same limit.

All the elements of $\mathbf{m}^{(T)}$ may be written in the form

$$m_i^{(T)} = \rho_i^{(T)} n_i^{(T)} \tag{4.10.11}$$

and it is clear that

$$\left.\begin{aligned} \rho_i^{(T)} &\geqslant r_T, \\ \rho_{L(T)}^{(T)} &= r_T, \end{aligned}\right\} \tag{4.10.12}$$

and

$$\rho_{U(T)}^{(T)} = R_T = r_T + d_T, \text{ say.} \tag{4.10.13}$$

These results are substituted into equations (4.10.5), and the first equation is then divided by the second. We obtain

$$\frac{m_i^{(T+1)}}{n_i^{(T+1)}} = \frac{\sum_{j=0}^{k} h_{ij}^{(T+1)}\{r_T + (\rho_j^{(T)} - r_T)\}\, n_j^{(T)}}{\sum_{j=0}^{k} h_{ij}^{(T+1)} n_j^{(T)}}$$

$$= r_T + \frac{\sum_{j=0}^{k} h_{ij}^{(T+1)}(\rho_j^{(T)} - r_T)\, n_j^{(T)}}{\sum_{j=0}^{k} h_{ij}^{(T+1)} n_j^{(T)}}$$

$$\geqslant r_T + d_T \left(\frac{h_{iU(T)}^{(T+1)} n_{U(T)}^{(T)}}{\sum_{j=0}^{k} h_{ij}^{(T+1)} n_j^{(T)}} \right). \quad (4.10.14)$$

We now examine the expression in brackets which for convenience we shall denote by η_{iT}. We note that

(i) η_{iT} is positive;

(ii) η_{iT} is less than one; and

(iii) η_{iT} is always bounded away from zero.

The first two results are immediately apparent, but the third requires some justification. It will be true if $n_i^{(T)}/n_j^{(T)}$ is bounded away from zero and infinity for all i, j and T, and this can be proved by induction. It is known from assumption (v) that the ratio

$$n_i^{(T)}/n_j^{(T)}$$

is bounded away from zero and infinity for $T = 0$. Assume that it is true for some T. Using inequality (4.10.4), we see that

$$\frac{n_i^{(T+1)}}{n_j^{(T+1)}} = \frac{\sum_{l=0}^{k} h_{il}^{(T+1)} n_l^{(T)}}{\sum_{l=0}^{k} h_{jl}^{(T+1)} n_l^{(T)}} > \frac{\epsilon^{N_0}}{M} = \frac{1}{R}, \text{ say.} \quad (4.10.15)$$

In a similar manner,

$$\frac{n_i^{(T+1)}}{n_j^{(T+1)}} < M\epsilon^{-N_0} = R. \quad (4.10.16)$$

Therefore (by induction) for all i, j and T, the ratios $\{n_i^{(T)}/n_j^{(T)}\}$ are bounded between two positive numbers R and $1/R$.

Substituting inequalities (4.10.15) and (4.10.16) into the definition of η_{iT}, we obtain

$$\eta_{iT} > \frac{h_{iU(T)}^{(T+1)}}{R\sum_{j=0}^{k} h_{ij}^{(T+1)}} > \frac{\epsilon^{N_0}}{R(k+1)M} = \frac{1}{(k+1)R^2} \qquad (4.10.17)$$

indicating the η_{iT} is always bounded away from zero.

If we define η_T equal to $\eta_{L(T+1)T}$, and make use of inequality (4.10.14), we obtain

$$r_{T+1} \geqslant r_T + \eta_T d_T, \qquad (4.10.18)$$

where $$\eta_T \geqslant \frac{1}{(k+1)R^2}. \qquad (4.10.19)$$

This argument which we have applied to the $\{r_T\}$ to derive equations (4.10.14) to (4.10.19) can also be applied to the $\{R_T\}$. We then obtain

$$R_{T+1} \leqslant R_T - \xi_T d_T \qquad (4.10.20)$$

where $$1 > \xi_T > \frac{1}{(k+1)R^2} > 0. \qquad (4.10.21)$$

The next step is to subtract inequality (4.10.18) from inequality (4.10.20):

$$(R_{T+1} - r_{T+1}) = d_{T+1} < d_T(1 - \xi_T - \eta_T). \qquad (4.10.22)$$

We deduce that

$$d_{T+1} \leqslant d_0 \prod_{i=0}^{T} (1 - \xi_i - \eta_i) \qquad (4.10.23)$$

which tends to zero as T tends to infinity. Finally

$$\lim_{T\to\infty} \left(\frac{m_i^{(T)}}{n_i^{(T)}} - \frac{m_j^{(T)}}{n_j^{(T)}} \right) = 0 \qquad (4.10.24)$$

for all i and all j. This completes the proof of the weak ergodicity result.

4.11 *Competing populations**

Some attempts have been made to use the linear approach of P. H. Leslie to study the growth of competing populations. These simple models are far removed from biological reality,[3] but they are not without their mathematical interest.

[3] The more realistic models are often studied by computer simulation (e.g. M. S. Bartlett, 1957).

Let us consider two distinct species of animals in the one (deterministic) environment. Both species depend upon the same food supply, and the animals therefore compete with members of their own species and with members of the other species for food. We shall ignore the problem of age structure, and denote the total number of animals of species 1 and 2 by $N_1(t)$ and $N_2(t)$ respectively. During time element $(t, t+1)$ there will be $b_1 N_1(t)$ births to species 1 and $b_2 N_2(t)$ births to species 2. There will also be $d_1 N_1(t)$ 'natural' deaths for species 1 and $d_2 N_2(t)$ 'natural' deaths for species 2.

Apart from the 'natural' deaths, there will be some animals that die as a direct result of competition. The number of deaths for species 1 and 2 due to competition will be assumed to be $c_{11} N_1(t) + c_{12} N_2(t)$ and $c_{21} N_1(t) + c_{22} N_2(t)$ respectively. Let us write a_{11} for $1 + b_1 - d_1 - c_{11}$, a_{12} for $-c_{12}$, a_{21} for $-c_{21}$ and a_{22} for $1 + b_2 - d_2 - c_{22}$. Then the following matrix recurrence equation will apply to the system

$$\begin{pmatrix} N_1(t+1) \\ N_2(t+1) \end{pmatrix} = \begin{pmatrix} a_{11} & a_{12} \\ a_{21} & a_{22} \end{pmatrix} \begin{pmatrix} N_1(t) \\ N_2(t) \end{pmatrix}. \tag{4.11.1}$$

An analysis of the ecological system will centre around the matrix in equation (4.11.1). It is important to remember that a_{12} and a_{21} are negative, and some care is necessary to avoid the possibility of a negative number of animals; this complication is introduced by the very simple assumptions made concerning the manner of competition.

The continuous-time version of this competition model is represented by the differential equation

$$\begin{pmatrix} N_1'(t) \\ N_2'(t) \end{pmatrix} = \begin{pmatrix} A_{11} & A_{12} \\ A_{21} & A_{22} \end{pmatrix} \begin{pmatrix} N_1(t) \\ N_2(t) \end{pmatrix}. \tag{4.11.2}$$

The diagonal elements of the matrix will normally be positive, and the off-diagonal elements will be negative. Equation (4.11.2) may be re-written as

$$\mathbf{N}'(t) = \mathbf{AN}(t). \tag{4.11.3}$$

To simplify the analysis, we shall assume that the two latent roots of $\mathbf{A}$ are distinct.[4] Then there exists a non-singular matrix $\mathbf{H}$ such that $\mathbf{H}^{-1}\mathbf{AH}$ is a diagonal matrix with non-zero elements λ_1 and λ_2 (the latent roots of $\mathbf{A}$). Let us make the transformation

$$\mathbf{n}(t) = \mathbf{H}^{-1}\mathbf{N}(t),$$

[4] If certain latent roots are repeated, minor complications are encountered like those mentioned in section 4.6. An exercise dealing with this problem is given in section 4.13.

in which case

$$\mathbf{n}'(t) = (\mathbf{H}^{-1}\mathbf{A}\mathbf{H})\,\mathbf{n}(t).$$

It follows that $n_1(t) \propto e^{\lambda_1 t}$ and $n_2(t) \propto e^{\lambda_2 t}$, and we conclude that

$$\left.\begin{aligned} N_1(t) &= \alpha_{11}e^{\lambda_1 t}+\alpha_{12}e^{\lambda_2 t};\\ N_2(t) &= \alpha_{21}e^{\lambda_1 t}+\alpha_{22}e^{\lambda_2 t}.\end{aligned}\right\} \qquad (4.11.4)$$

There is no theoretical difficulty in introducing an age structure into the discrete-time competitive model, and both the discrete- and continuous-time models can be extended to deal with three or more species in competition. Simple linear predator–prey models can be developed in the same manner.

4.12 *An example*

This interesting example is due to R. MacArthur (1968), and it involves a simple population with two age groups.

$$\begin{pmatrix} n_{0,t+1}\\ n_{1,t+1}\end{pmatrix} = \begin{pmatrix} F_0 & F_1\\ P_0 & 0\end{pmatrix}\begin{pmatrix} n_{0,t}\\ n_{1,t}\end{pmatrix}. \qquad (4.12.1)$$

In another time interval, the recurrence matrix is altered so that F_0 becomes F_0', F_1 becomes F_1' and P_0 becomes P_0'. Then if the primed years alternate with the unprimed ones,

$$\begin{pmatrix} n_{0,t+2}\\ n_{1,t+2}\end{pmatrix} = \begin{pmatrix} F_0'F_0+P_0F_1' & F_0'F_1\\ P_0'F_0 & P_0'F_1\end{pmatrix}\begin{pmatrix} n_{0,t}\\ n_{1,t}\end{pmatrix}. \qquad (4.12.2)$$

Every second time unit, the environment repeats the cycle, and we are interested in the intrinsic rate of increase of the population. The characteristic equation of the product matrix is

$$\lambda^2-(F_1'P_0+F_1P_0'+F_0F_0')\,\lambda+F_1F_1'P_0P_0' = 0,$$

so that the latent roots of the product matrix are given by

$$\lambda = \tfrac{1}{2}[(F_1'P_0+F_1P_0')+F_0F_0' \pm \{(F_1'P_0-F_1P_0')^2+ \\ +(F_0F_0')^2+2F_0F_0'(F_1P_0'+F_1'P_0)\}^{\frac{1}{2}}]. \qquad (4.12.3)$$

The following points should now be noted:

(i) The terms in $(F_1'P_0+F_1P_0')$ are unaltered if $F_1'P_0$ and F_1P_0' are set equal to their average value.

(ii) The term $(F_1'P_0-F_1P_0')^2$ is larger when $F_1'P_0$ is different from F_1P_0'; an increase in the difference will cause an increase in the larger latent root.

(iii) $(\bar{F}_0 - z)(\bar{F}_0 + z) = \bar{F}_0^2 - z^2$, so that the larger latent root is reduced when the average value of $\bar{F}_0$ is replaced by $(\bar{F}_0 + z)$ and $(\bar{F}_0 - z)$ in alternate years.

If for example, the fertility rates are kept constant and the survivorship probability from age group 0 to age group 1 alternates between P_0 and P_0', the population has a greater intrinsic rate of natural increase than when the survival probability remains fixed at $\frac{1}{2}(P_0 + P_0')$. The effect of an oscillating birth rate to mothers aged 0 is to reduce the intrinsic rate of natural increase, other rates being kept constant.

This type of model may be appropriate to certain insects which reproduce during two different seasons each year. The environment conditions will be considerably different in the two seasons. Generalizations of these results clearly exist, but the formulation of the more general problem is quite difficult, and no results have yet been proven.

The periodicity of the oscillation is important. In the more general formulation, the period of oscillation must approximate the generation time of the population. When the fluctuations are random, the results are no longer valid. The case of random fluctuations is considered later in section 9.8.

4.13 *Exercises*

1 'Take a species, say a beetle, which lives three years only, and which propagates in its third year of life. Let the survival rate of the first age-group be $\frac{1}{2}$, of the second $\frac{1}{3}$, and assume that each female in the age 2–3 produces, in the average, 6 new living females.' (H. Bernardelli, 1941.) Analyse this population assuming a thousand females in each age group at time $t = 0$.

2 The following two Leslie matrices apply to certain animal populations:

$$\begin{pmatrix} 0 & 0 & 0 & 14 & 0 & 20 \\ \frac{4}{5} & & & & & \\ & \frac{3}{4} & & & & \\ & & \frac{5}{6} & & & \\ & & & \frac{4}{5} & & \\ & & & & \frac{3}{4} & 0 \end{pmatrix}, \quad \begin{pmatrix} 0 & 0 & 14 & 0 & 20 \\ \frac{4}{5} & & & & \\ & \frac{3}{4} & & & \\ & & \frac{5}{6} & & \\ & & & \frac{4}{5} & 0 \end{pmatrix}$$

Are these matrices positive regular?

3 Determine all the latent roots of the first matrix in question 2.

4 A certain population of insects will survive and reproduce in one environment according to the first Leslie matrix exhibited below and, in another environment, the population will obey the second Leslie matrix. Prove that the population latent roots are identical in the two different environments.

$$\begin{pmatrix} 0 & 2 & 3 \\ \frac{1}{2} & 0 & 0 \\ 0 & \frac{2}{3} & 0 \end{pmatrix}; \quad \begin{pmatrix} 0 & 3 & 4 \\ \frac{1}{3} & 0 & 0 \\ 0 & \frac{3}{4} & 0 \end{pmatrix}.$$

5 The population described in question 4 is placed in an environment which oscillates from one point of time to the next between the two environments described in question 4. Will the population increase at the same intrinsic rate, at a faster intrinsic rate or at a slower intrinsic rate?

6 A positive regular Leslie matrix has *exactly one* positive latent root. The theorem of Perron and Frobenius tells us that a positive regular matrix has a positive latent root which is dominant, but the theorem does not preclude further positive roots. Construct a 2×2 example with two positive roots.

7 Find the inverse of the Leslie matrix and suggest a possible use for it.

8 Describe the latent roots and latent vectors of $\mathbf{A}^{-1}$ in terms of similar quantities pertaining to $\mathbf{A}$. Is it possible to say anything about a dominant latent root for $\mathbf{A}^{-1}$?

9 An animal population is growing in an oscillating environment like that described in question 5, and the two environments correspond to the summer six months and the winter six months. The survival rates and fertility measures in summer and winter are those exhibited in matrices one and two respectively of question 4, the time unit being six months. The number of these animals is rapidly approaching plague proportions, and the government is about to initiate an extermination programme. Unfortunately the method involved is rather expensive. It is believed that the scheme will kill 50% of the animals aged under 6 months, $66\frac{2}{3}$% of all animals aged 6–12 months, and 75% of the animals aged over 12 months within a few days of each annual application. Should the scheme be applied annually in Autumn or Spring?

10 A demographer intends to project the Australian female population for a particular purpose and he wishes to exclude future immigrants. He believes that the Leslie matrix method is suitable for his purpose and he intends using a time and age interval of one year. Suggest a suitable method for determining the $\{P_j\}$ values.

11 The values $\{F_j\}$ $(j = 0, 1, \ldots, k)$ in a certain population all increase by 1%.
(i) Give a formula for the change in λ_0.
(ii) How is the net reproduction rate R_0 affected?
(iii) Reconcile these results with formula (3.5.1).

12 The following continuous-time matrix $\mathbf{A}$ has been suggested as being suitable for a predator–prey model:

$$\begin{pmatrix} 0.40 & 0.05 \\ -0.20 & 0.60 \end{pmatrix}.$$

Obtain formulae for $N_1(t)$ and $N_2(t)$, assuming that $N_1(0) = 100$ and $N_2(0) = 100$.

13 All the elements in the first row of a 15×15 Leslie matrix are zero except F_5, F_9 and F_{14}. Is the matrix positive regular?

5
Simple birth and death processes

5.1 *Introduction*

All the models for human populations developed before 1946 were both deterministic and unisexual and the problem of producing a useful stochastic model had to wait until the study of stochastic processes had developed sufficiently. Many of the early stochastic processes were developed in connection with bacterial populations, particle physics and telephone exchange problems. They are conveniently referred to as birth and death processes, and although they are unlikely to find much practical application with human populations, it is instructive to study a few simple examples because the models involved have influenced the development of stochastic models for human populations. The early work in this field was done by A. G. McKendrick (1926), G. U. Yule (1924), W. Feller (1939) and D. G. Kendall (1949). The paper by Kendall contains a wealth of information and we shall refer to it on numerous occasions throughout this book.

5.2 *The Poisson process*

Consider a population subject to births but no deaths. The probability of a single birth in time element $(t, t+\mathrm{d}t)$ will be assumed to be

$$\lambda\,\mathrm{d}t + o(\mathrm{d}t), \tag{5.2.1}$$

where λ is independent of the size and age-structure of the population and its previous history, and also independent of time. The probability of zero births in the time element is $1-\lambda\,\mathrm{d}t+o(\mathrm{d}t)$ and the probability of two or more births is of smaller order than $\mathrm{d}t$. We shall assume also that the population has no members at time $t=0$. Then if $P_n(t)$ denotes the probability that the population is made up of n individuals at time t, it is soon apparent that

$$\left.\begin{aligned} P_0(t+\mathrm{d}t) &= P_0(t)\,(1-\lambda\,\mathrm{d}t)+o(\mathrm{d}t);\\ P_n(t+\mathrm{d}t) &= P_n(t)\,(1-\lambda\,\mathrm{d}t)+P_{n-1}(t)\,\lambda\,\mathrm{d}t+o(\mathrm{d}t). \end{aligned}\right\} \tag{5.2.2}$$

That is,

$$\left.\begin{aligned} \frac{P_0(t+\mathrm{d}t)-P_0(t)}{\mathrm{d}t} &= -\lambda P_0(t)+o(1); \\ \frac{P_n(t+\mathrm{d}t)-P_n(t)}{\mathrm{d}t} &= -\lambda P_n(t)+\lambda P_{n-1}(t)+o(1); \end{aligned}\right\} \tag{5.2.3}$$

whence

$$\left.\begin{aligned} P_0'(t) &= -\lambda P_0(t); \\ P_n'(t) &= -\lambda P_n(t)+\lambda P_{n-1}(t). \end{aligned}\right\} \tag{5.2.4}$$

The dash denotes a derivative with respect to t.

From the first of the two equations (5.2.4) and the fact that $P_0(0) = 1$, it is clear that

$$P_0(t) = \mathrm{e}^{-\lambda t}. \tag{5.2.5}$$

This result may be substituted in the second of the pair of equations (5.2.4) to yield an inhomogeneous first-order differential equation for $P_1(t)$:

$$P_1'(t)+\lambda P_1(t) = \lambda\,\mathrm{e}^{-\lambda t}. \tag{5.2.6}$$

The integrating factor for this differential equation is $\mathrm{e}^{\lambda t}$, so that we obtain

$$\frac{\mathrm{d}}{\mathrm{d}t}\{P_1(t)\,\mathrm{e}^{\lambda t}\} = \lambda. \tag{5.2.7}$$

Using the initial condition $P_1(0) = 0$, we find that

$$P_1(t) = (\lambda t)\,\mathrm{e}^{-\lambda t}. \tag{5.2.8}$$

It may be proved by induction that

$$P_n(t) = \frac{(\lambda t)^n}{n!}\,\mathrm{e}^{-\lambda t}. \tag{5.2.9}$$

The population size at time t is therefore a Poisson random variable with mean λt.

5.3 *The Yule process*

With the population of section 5.2, the probability of a birth in time element $(t, t+\mathrm{d}t)$ is independent of the population size. In most biological contexts, and certainly in demographic contexts, this assumption is very unrealistic. We are therefore led to the study of a population in which the probability of a birth in the time element $(t, t+\mathrm{d}t)$ is proportional to the population size n at time t. Each individual member of this population is assumed to act independently and give birth to a new individual during time element dt with probability $\lambda\,\mathrm{d}t$. For the population as a whole therefore

$$\begin{aligned}\Pr(\text{exactly one birth in time } \mathrm{d}t) &= n\lambda\,\mathrm{d}t + o(\mathrm{d}t);\\ \Pr(\text{zero births in time } \mathrm{d}t) &= 1 - n\lambda\,\mathrm{d}t + o(\mathrm{d}t);\\ \Pr(\text{two or more births in time } \mathrm{d}t) &= o(\mathrm{d}t).\end{aligned}$$

We shall assume an initial population size of one. It is soon apparent that

$$P_n(t+\mathrm{d}t) = P_n(t)\,(1 - n\lambda\,\mathrm{d}t) + P_{n-1}(t)\,\lambda(n-1)\,\mathrm{d}t + o(\mathrm{d}t), \quad (5.3.1)$$

for $n \geqslant 1$, and $P_0(t) = 0$ for all t.

Using the techniques of section 5.2, it may be shown that

$$P_n(t) = \mathrm{e}^{-\lambda t}(1 - \mathrm{e}^{-\lambda t})^{n-1}. \quad (5.3.2)$$

The case in which the population is of size m initially is set as an exercise.

5.4 *A linear birth and death process*

Let us now modify the model of section 5.3 and assume that each individual existing at time t has a chance

$$\mu\,\mathrm{d}t + o(\mathrm{d}t) \quad (5.4.1)$$

of dying in the following time interval of length dt. The force of mortality μ is the same for all individuals at all times t. Then

$$\left.\begin{aligned}P_0(t+\mathrm{d}t) &= P_0(t) + P_1(t)\,\mu\,\mathrm{d}t + o(\mathrm{d}t);\\ P_n(t+\mathrm{d}t) &= P_{n-1}(t)\,(n-1)\,\lambda\,\mathrm{d}t + P_{n+1}(t)\,(n+1)\,\mu\,\mathrm{d}t\\ &\quad + P_n(t)\,\{1 - n(\lambda+\mu)\,\mathrm{d}t\} + o(\mathrm{d}t) \quad (n \geqslant 1);\end{aligned}\right\} \quad (5.4.2)$$

and

$$\left.\begin{aligned}P_0'(t) &= \mu P_1(t);\\ P_n'(t) &= (n-1)\,\lambda P_{n-1}(t) + (n+1)\,\mu P_{n+1}(t)\\ &\quad - n(\lambda+\mu)\,P_n(t) \quad (n \geqslant 1).\end{aligned}\right\} \quad (5.4.3)$$

It is rather more difficult to obtain an explicit solution for this stochastic process. The expectation and variance of the population size $N(t)$ at time t may be derived in a simple manner however. Consider the equations (5.4.3) multiplied throughout by n and summed for all non-negative values of n:

$$\begin{aligned}\frac{\mathrm{d}}{\mathrm{d}t}\sum_{n=0}^{\infty} nP_n(t) &= \sum_{n=0}^{\infty}(n+1)^2\mu P_{n+1}(t) - \sum_{n=0}^{\infty}(n+1)\,\mu P_{n+1}(t)\\ &\quad + \sum_{n=1}^{\infty}(n-1)^2\lambda P_{n-1}(t) + \sum_{n=1}^{\infty}(n-1)\,\lambda P_{n-1}(t)\\ &\quad - \sum_{n=0}^{\infty} n^2(\lambda+\mu)\,P_n(t).\end{aligned}$$

This equation simplifies to

$$\frac{d}{dt} M_1(t) = (\lambda - \mu) M_1(t), \tag{5.4.4}$$

where $M_1(t)$ is the expected value of $N(t)$. It is clear that the expected population will follow the Malthusian growth pattern:

$$M_1(t) = e^{(\lambda-\mu)t}, \tag{5.4.5}$$

(assuming a single initial ancestor).

An expression for the variance of the population size at time t may be derived in a similar manner. We find that

$$\frac{d}{dt} M_2(t) - 2(\lambda - \mu) M_2(t) = (\lambda + \mu) M_1(t), \tag{5.4.6}$$

where $M_2(t)$ is the second moment of $N(t)$ about the origin. It is then easy to prove that

$$\text{Var } N(t) = \frac{\lambda+\mu}{\lambda-\mu} e^{(\lambda-\mu)t} \{e^{(\lambda-\mu)t} - 1\}. \tag{5.4.7}$$

This formula is due to W. Feller (1939). Again, we have assumed a single initial ancestor. We also assume that $\lambda \neq \mu$.

Let us now return to the problem of an explicit solution to equations (5.4.3), and define a probability-generating function

$$\phi(z, t) = \sum_{n=0}^{\infty} P_n(t) z^n. \tag{5.4.8}$$

Equations (5.4.3) may be multiplied throughout by z^n and summed for all non-negative values of n to yield

$$\frac{\partial \phi}{\partial t} = (\lambda z - \mu)(z - 1) \frac{\partial \phi}{\partial z}. \tag{5.4.9}$$

We require ϕ as a function of z and t.

Consider a contour of $\phi(z, t)$, $z = z(t)$ say, for which ϕ is constant. Then

$$\frac{d\phi}{dz} = \frac{\partial \phi}{\partial z} + \frac{\partial \phi}{\partial t}\frac{dt}{dz} = 0.$$

On this contour therefore,

$$\frac{dt}{dz} = -\frac{\partial \phi}{\partial z}\left(\frac{\partial \phi}{\partial t}\right)^{-1} \tag{5.4.10}$$

If this result is substituted into equation (5.4.9), we know that

$$dt = \frac{-dz}{(\lambda z - \mu)(z - 1)} \tag{5.4.11}$$

on the contour, whence

$$\frac{\lambda z-\mu}{z-1}\mathrm{e}^{-(\lambda-\mu)t} = C \tag{5.4.12}$$

provided $\lambda \neq \mu$. The constant C depends upon the contour selected and it is therefore a function of the height of the contour above the (z, t) plane.

If we select an arbitrary point (z_0, t_0) in the (z, t) plane, the $\phi(z, t)$ contour above this point will obey the equations

$$C_0 = \frac{\lambda z_0-\mu}{z_0-1}\mathrm{e}^{-(\lambda-\mu)t_0} = \frac{\lambda z-\mu}{z-1}\mathrm{e}^{-(\lambda-\mu)t},$$

and the height of the contour will be $\phi(z_0, t_0)$. The constant C_0 will be a function of $\phi(z_0, t_0)$, and it is therefore possible to write

$$C_0 = F(\phi(z_0, t_0)). \tag{5.4.13}$$

This relationship is true for every point (z_0, t_0) in the appropriate area of the (z, t) plane. In particular, it will be true for the point $(z, 0)$, and we then have

$$\frac{\lambda z-\mu}{z-1} = F(z),$$

since $$\phi(z, 0) = z, \tag{5.4.14}$$

assuming a single initial ancestor. We note that

$$F^{-1}(y) = \frac{\mu-y}{\lambda-y}$$

and conclude from equation (5.4.13) that

$$\phi(z_0, t_0) = F^{-1}(C_0) = F^{-1}\left(\left(\frac{\lambda z_0-\mu}{z_0-1}\right)\mathrm{e}^{-(\lambda-\mu)t_0}\right).$$

Finally therefore,

$$\phi(z, t) = \frac{\mu-\dfrac{\lambda z-\mu}{z-1}\mathrm{e}^{-(\lambda-\mu)t}}{\lambda-\dfrac{\lambda z-\mu}{z-1}\mathrm{e}^{-(\lambda-\mu)t}}. \tag{5.4.15}$$

This probability-generating function describes fully the distribution of the population size at time t. It is possible to determine the moments of the distribution by differentiating $\phi(z, t)$ with respect to z and setting z equal to one. Formulae (5.4.5) and (5.4.7) could be verified in this manner.

Equation (5.4.15) enables us to investigate the probability of ultimate extinction of the population. It is clear that this probability is given by

$$\lim_{t\to\infty} P_0(t) = \lim_{t\to\infty} \phi(0, t), \tag{5.4.16}$$

and it is soon apparent that the population becomes extinct with probability one if $\mu > \lambda$, and with probability μ/λ if $\lambda > \mu$. The case $\lambda = \mu$ is set as an exercise.

5.5 *Kendall's birth, death and migration model*

D. G. Kendall (1948*a*) suggested a modification of the linear birth and death process of section 5.4 to include immigration. He assumed that the population would be increased by a single immigrant in time element dt with probability

$$\kappa \mathrm{d}t + o(\mathrm{d}t). \tag{5.5.1}$$

This assumption leads us to the following differential-difference equation:

$$P_n'(t) = \{(n-1)\lambda + \kappa\} P_{n-1}(t) + (n+1)\mu P_{n+1}(t) - \{n(\lambda+\mu)+\kappa\} P_n(t). \tag{5.5.2}$$

The methods of section 5.4 are again applicable.

To determine the mean population size at time t, we multiply equation (5.5.2) by n and sum from $n = 0$ to infinity, and the following intuitively obvious first-order differential equation is obtained:

$$\frac{\mathrm{d}}{\mathrm{d}t} M_1(t) = (\lambda - \mu) M_1(t) + \kappa. \tag{5.5.3}$$

If we assume that $\lambda \neq \mu$, the integrating factor for the differential equation is $e^{-(\lambda-\mu)t}$, and it follows that

$$M_1(t) = \frac{\kappa}{\lambda - \mu}\{e^{(\lambda-\mu)t} - 1\} \tag{5.5.4}$$

if the initial population has zero members. The determination of the variance is set as an exercise.

To determine the distribution of the population size at time t, we again make use of a probability-generating function $\phi(z, t)$. Equation (5.5.2) is multiplied by z^n and summed for all non-negative n to obtain the following differential equation:

$$\frac{\partial \phi}{\partial t} = (\lambda z - \mu)(z-1)\frac{\partial \phi}{\partial z} + \kappa(z-1)\phi. \tag{5.5.5}$$

In solving this equation, which is a partial differential equation of the standard Lagrangian type (I. N. Sneddon, 1957), we seek a relationship of the form

$$F(z, t, \phi) = 0. \tag{5.5.6}$$

In the search for this function, let us consider first the parametric representation of a curve in the (z, t, ϕ) space:

$$\left.\begin{aligned} z &= z(s), \\ t &= t(s), \\ \phi &= \phi(s), \end{aligned}\right\} \tag{5.5.7}$$

where the parameter s has been chosen as the distance along the curve from a certain fixed point. It is soon apparent that the direction cosines of the tangent to the curve at the point (z, t, ϕ) are

$$\left(\frac{dz}{ds}, \frac{dt}{ds}, \frac{d\phi}{ds}\right). \tag{5.5.8}$$

The curve we have described so far is completely arbitrary. Let us now assume that it is constrained to lie in the surface defined by equation (5.5.6). Then

$$F(z(s), t(s), \phi(s)) = 0, \tag{5.5.9}$$

and
$$\frac{\partial F}{\partial z}\frac{dz}{ds} + \frac{\partial F}{\partial t}\frac{dt}{ds} + \frac{\partial F}{\partial \phi}\frac{d\phi}{ds} = 0. \tag{5.5.10}$$

Equation (5.5.10) in conjunction with result (5.5.8) indicates that *every* curve passing through the point (z, t, ϕ) on the surface must be perpendicular to a line whose direction cosines are proportional to

$$\left(\frac{\partial F}{\partial z}, \frac{\partial F}{\partial t}, \frac{\partial F}{\partial \phi}\right). \tag{5.5.11}$$

This line is of course the normal to the surface at that point.

Equation (5.5.6) may be written in the slightly different form

$$\phi - \phi(z, t) = 0, \tag{5.5.12}$$

in which case the direction cosines of the normal to the surface are proportional to

$$\left(\frac{\partial \phi}{\partial z}, \frac{\partial \phi}{\partial t}, -1\right). \tag{5.5.13}$$

Now, according to equation (5.5.5),

$$(\lambda z - \mu)(z - 1)\frac{\partial \phi}{\partial z} - \frac{\partial \phi}{\partial t} + \kappa(z - 1)\phi = 0, \tag{5.5.14}$$

and we conclude that a line passing through the point (z, t, ϕ) with its direction cosines proportional to

$$((\lambda z-\mu)(z-1),\ -1,\ -\kappa(z-1)\phi) \tag{5.5.15}$$

must be tangential to the surface.

If we start at an arbitrary point in the surface and move in such a manner that the direction of motion is always according to the direction cosines given by (5.5.15) at the current point, we trace out a curve on the surface. The functions in (5.5.15) are unique, so at each point in the surface there is exactly one such curve passing through the point.

Consider a particular point on the surface, and let the unique curve through that point be defined parametrically by equations (5.5.7). The direction cosines of the curve are given in (5.5.8) and these must be proportional to those in (5.5.15). Therefore

$$\frac{dz}{(\lambda z-\mu)(z-1)} = \frac{dt}{-1} = \frac{d\phi}{-\kappa(z-1)\phi}. \tag{5.5.16}$$

Equations (5.5.16) have solutions

$$\left.\begin{aligned}\frac{\lambda z-\mu}{z-1}e^{-(\lambda-\mu)t} &= C,\\ (\lambda z-\mu)\phi^{\lambda/\kappa} &= K.\end{aligned}\right\} \tag{5.5.17}$$

Both C and K depend upon the particular curve under consideration, and it follows that K must be a function of C. A general solution to equation (5.5.5) is therefore of the form

$$(\lambda z-\mu)\phi^{\lambda/\kappa} = \Phi\left(\left(\frac{\lambda z-\mu}{z-1}\right)e^{-(\lambda-\mu)t}\right). \tag{5.5.18}$$

At time $t = 0$, the population is assumed to have zero members, so that $\phi(z, 0) = 1$. From equation (5.5.18) therefore

$$(\lambda z-\mu) = \Phi\left(\frac{\lambda z-\mu}{z-1}\right). \tag{5.5.19}$$

Writing y instead of $(\lambda z-\mu)/(z-1)$, we see that

$$\Phi(y) = \frac{y(\lambda-\mu)}{y-\lambda}.$$

We deduce from equation (5.5.18) that

$$\phi(z,t) = \left\{\frac{\lambda-\mu}{\lambda e^{(\lambda-\mu)t}-\mu}\right\}^{\kappa/\lambda}\left\{1-\lambda z\frac{e^{(\lambda-\mu)t}-1}{\lambda e^{(\lambda-\mu)t}-\mu}\right\}^{-\kappa/\lambda}. \tag{5.5.20}$$

This probability-generating function defines the distribution of the population size at time t.

5.6 *An example*

A certain population is governed by the laws of Kendall's birth, death and immigration model. It is known that λ is strictly less than μ and that κ is positive. Under these circumstances it is soon apparent that the distribution of the population size is asymptotically independent of t as t tends to infinity. Find the stationary distribution.

Solution. Consider equation (5.5.2). When the stationary condition is attained, $P_n(t)$ will be independent of t, and its derivative with respect to t will be zero. Then

$$\{(n-1)\lambda+\kappa\}P_{n-1}+(n+1)\mu P_{n+1}-\{n(\lambda+\mu)+\kappa\}P_n = 0. \tag{5.6.1}$$

Multiplying by z^n and summing for all non-negative values of n, we obtain the differential equation

$$(\lambda z-\mu)(z-1)\frac{d\phi}{dz}+\kappa(z-1)\phi = 0,$$

where $\phi(z)$ is the probability-generating function of the stationary distribution. We see that

$$\frac{d\phi}{\phi} = \frac{-\kappa dz}{\lambda z-\mu},$$

and deduce that

$$\phi = C(\lambda z-\mu)^{-\kappa/\lambda}.$$

The constant C is chosen such that the probabilities sum to one. Then

$$\phi = \{(1-\lambda/\mu)/(1-\lambda z/\mu)\}^{\kappa/\lambda}. \tag{5.6.2}$$

This formula can also be derived from equation (5.5.20) by considering the limit at $t\to\infty$.

If κ/λ is an integer, this function is the probability-generating function of a negative binomial distribution. If κ/λ is very small, the limiting conditional distribution of the population size $N(t)$ (given that $N(t) > 0$) is the Fisher logarithmic series distribution (D. G. Kendall, 1948*a*).

5.7 *Exercises*

1 Use the method of the text to derive $P_3(t)$ from $P_2(t)$ for the Yule process.

2 Derive the mean and variance of the Yule process using the method described in the second paragraph of section 5.4.

3 Derive the mean and variance for the Yule process using equation (5.3.2).

4 Derive equations (5.4.6) and (5.4.7) for the linear birth and death model when $\lambda \neq \mu$.

5 Derive the equation corresponding to (5.4.7) for the linear birth and death model when $\lambda = \mu$.

6 Solve equation (5.4.9) when $\lambda = \mu$, assuming a single initial ancestor.

7 Use the results of question 6 to verify the variance determined in question 5.

8 What is the probability of ultimate extinction for the linear birth and death model when $\lambda = \mu$?

9 Determine the variance of the population size at time t for Kendall's birth, death and immigration model when $\lambda \neq \mu$.

10 Solve equation (5.5.5) for the case $\lambda = \mu$, assuming a zero initial population.

11 Determine the mean and variance of Kendall's population when $\lambda = \mu$.

12 What is the probability of ultimate extinction for Kendall's population?

13 Derive equation (5.6.2) from formula (5.5.20).

14 A population is governed by the laws of growth of Kendall's birth, death and immigration model. The parameter κ is small and λ is less than μ. Prove that the conditional distribution of the population size given that it is non-zero follows Fisher's logarithmic series distribution:

$$\Pr\{N(t) = j \mid N(t) \neq 0\} = \left\{-\log\left(1-\frac{\lambda}{\mu}\right)\right\}^{-1} \frac{(\lambda/\mu)^j}{j} \quad (j = 1, 2, 3, \ldots).$$

15 For a population of the Kendall type, $\lambda = 1$, $\mu = 2$ and $\kappa = 1$. Find the probability that the population has zero members at time t. What is the limit as t tends to infinity?

16 Use formula (5.6.2) to find the probability that the Kendall population of question 15 has zero members at time t where t is large.

17 A population of the Yule type is of size m at time $t = 0$. Derive a formula for $P_n(t)$.

18 What is the probability of extinction for a linear birth and death population having N_0 members at time $t = 0$?

6

The stochastic models of M. S. Bartlett and D. G. Kendall*

6.1 *Introduction*

A few isolated stochastic processes were studied prior to 1940. During the following decade, however, many such processes were investigated and the literature is now extensive. M. S. Bartlett gave a course on stochastic processes at the University of North Carolina during the Fall Quarter of 1946, and in that course he included several topics of direct interest to population mathematicians. Two of these topics are discussed in sections 6.2 and 6.3.

Bartlett's methods involve discrete time and discrete age groups and his analysis was designed to yield asymptotic forms for the linear and quadratic moments when the age groups and time-intervals were both made small. From this point of view we may say that Bartlett's work lies half-way between a discrete and a continuous analysis. His work inspired D. G. Kendall (1949) to develop a model involving continuous-time and a continuous age structure. Kendall's analysis is described in sections 6.4 and 6.5.

6.2 *A two-type example due to Bartlett*

Suppose the deterministic population growth equation for a simple organism takes the form

$$\begin{pmatrix} n_{0,t+1} \\ n_{1,t+1} \end{pmatrix} = \begin{pmatrix} 0 & 2 \\ \frac{1}{2} & 0 \end{pmatrix} \begin{pmatrix} n_{0,t} \\ n_{1,t} \end{pmatrix} \tag{6.2.1}$$

corresponding to an expectation of two offspring in the second age group, and a chance of survival from the first to the second age group of $\frac{1}{2}$. Such a population is obviously non-increasing on average, and the solution for the case in which $n_{0,0}$ and $n_{1,0}$ are both equal to $\frac{1}{2}n$ is easily found to be

$$\begin{pmatrix} n_{0,t} \\ n_{1,t} \end{pmatrix} = \tfrac{3}{4}n \begin{pmatrix} 1 \\ \frac{1}{2} \end{pmatrix} - \tfrac{1}{4}n(-1)^t \begin{pmatrix} 1 \\ -\frac{1}{2} \end{pmatrix}. \tag{6.2.2}$$

Suppose now, for the complete *stochastic* model that the number of offspring is indeed fixed at two, but that the chance of survival to the reproductive age is an independent chance of $\frac{1}{2}$ for each individual in age group 0. The quantities $n_{0,t}$ and $n_{1,t}$ are now random variables, and their joint moment-generating function at time $t+1$ is

$$M_{t+1}(\theta_0, \theta_1) = \mathscr{E} \exp(\theta_0 n_{0,t+1} + \theta_1 n_{1,t+1}).$$

But for this model $n_{0,t+1} = 2n_{1,t}$, and $n_{1,t+1}$ is a conditional binomial random variable $B(n_{0,t}, \frac{1}{2})$ conditional on $n_{0,t}$. The moment-generating function of a binomial random variable $B(n, p)$ is $(q + p\mathrm{e}^{\theta})^n$, and so

$$\begin{aligned} M_{t+1}(\theta_0, \theta_1) &= \mathscr{E}(\mathrm{e}^{2n_{1,t}\theta_0})\,(\tfrac{1}{2}+\tfrac{1}{2}\mathrm{e}^{\theta_1})^{n_{0,t}} \\ &= \mathscr{E} \exp\{2n_{1,t}\theta_0 + n_{0,t}\log(\tfrac{1}{2}+\tfrac{1}{2}\mathrm{e}^{\theta_1})\} \\ &= M_t(\log(\tfrac{1}{2}+\tfrac{1}{2}\mathrm{e}^{\theta_1}), 2\theta_0) \end{aligned} \qquad (6.2.3)$$

a functional equation for M_t.

Now $K_t(\theta_0, \theta_1) = \log M_t(\theta_0, \theta_1)$ is the cumulant-generating function for the process, and clearly

$$K_{t+1}(\theta_0, \theta_1) = K_t(\log(\tfrac{1}{2}+\tfrac{1}{2}\mathrm{e}^{\theta_1}), 2\theta_0). \qquad (6.2.4)$$

Let us denote the expected value of $n_{j,t}$ by $e_{j,t}$, the variance of $n_{j,t}$ by $C_{j,j}{}^{(t)}$, and the covariance $\mathrm{Cov}(n_{i,t}, n_{j,t})$ by $C_{i,j}{}^{(t)}$. Equation (6.2.4) may be expanded in powers of θ_0 and θ_1, and we then equate the coefficients of the powers of θ_0 and θ_1 of first and second order.[1] The following result is obtained.

$$\begin{pmatrix} e_{0,t+1} \\ e_{1,t+1} \\ \hline C_{0,0}{}^{(t+1)} \\ C_{0,1}{}^{(t+1)} \\ C_{1,0}{}^{(t+1)} \\ C_{1,1}{}^{(t+1)} \end{pmatrix} = \left(\begin{array}{cc|cccc} 0 & 2 & 0 & 0 & 0 & 0 \\ \frac{1}{2} & 0 & 0 & 0 & 0 & 0 \\ \hline 0 & 0 & 0 & 0 & 0 & 4 \\ 0 & 0 & 0 & 0 & 1 & 0 \\ 0 & 0 & 0 & 1 & 0 & 0 \\ \frac{1}{4} & 0 & \frac{1}{4} & 0 & 0 & 0 \end{array}\right) \begin{pmatrix} e_{0,t} \\ e_{1,t} \\ \hline C_{0,0}{}^{(t)} \\ C_{0,1}{}^{(t)} \\ C_{1,0}{}^{(t)} \\ C_{1,1}{}^{(t)} \end{pmatrix}. \qquad (6.2.5)$$

This is a linear recurrence relation connecting the expectations and second-order central moments at times t and $t+1$. The notation and form of equation are due to J. H. Pollard (1966).

From equation (6.2.5) it may be noted that

$$C_{0,0}{}^{(t+2)} = 4C_{1,1}{}^{(t+1)} = C_{0,0}{}^{(t)} + e_{0,t} = C_{0,0}{}^{(t-2)} + e_{0,t} + e_{0,t-2}.$$

But, from equation (6.2.2) we know that $e_{0,t-2r}$ is constant and positive for all integral r. Therefore the variance $C_{0,0}{}^{(t)}$ increases indefinitely as t increases. It follows that the expectations give little information

[1] The use of the cumulant-generating function is described by M. G. Kendall and A. Stuart (1963), 67–89.

about the process for large t. Indeed, it is possible to prove that the process becomes extinct with probability one.

If we substitute z_0 and z_1 for e^{θ_0} and e^{θ_1} respectively, equation (6.2.3) becomes

$$M_{t+1}(\log z_0, \log z_1) = M_t(\log(\tfrac{1}{2}+\tfrac{1}{2}z_1), \log(z_0^2)).$$

But the expected value of $z_0^{n_{0,t}}z_1^{n_{1,t}}$ is the probability-generating function $\phi_t(z_0, z_1)$ for the bivariate distribution of $n_{0,t}$ and $n_{1,t}$, and we conclude that

$$\phi_{t+1}(z_0, z_1) = \phi_t(\tfrac{1}{2}+\tfrac{1}{2}z_1, z_0^2). \tag{6.2.6}$$

Applying this equation recursively we find that

$$\phi_{t+2r}(0, 0) = \phi_{t+2r-2}(\tfrac{1}{2}, (\tfrac{1}{2})^2) = \phi_{t+2r-4}(\tfrac{1}{2}+\tfrac{1}{2}(\tfrac{1}{2})^2, \{\tfrac{1}{2}+\tfrac{1}{2}(\tfrac{1}{2})^2\}^2)$$

and so on. The limit of the steadily increasing sequence

$$u_1 = \tfrac{1}{2}+\tfrac{1}{2}(\tfrac{1}{2})^2, \quad u_2 = \tfrac{1}{2}+\tfrac{1}{2}u_1^2, \quad u_3 = \tfrac{1}{2}+\tfrac{1}{2}u_2^2, \quad \ldots$$

is given by the solution of the equation $x = \frac{1}{2}+\frac{1}{2}x^2$; i.e. $x = 1$. Therefore

$$\lim_{r\to\infty} \phi_{t+2r}(0, 0) = \phi_t(1, 1),$$

which is equal to one. That is, the probability of ultimate extinction is one. This result shows that it cannot be assumed that the mean in a finite population undergoing stochastic changes remains representative of the actual numbers that exist after a long time.

This simple two-type process is an example of a stochastic process usually called a *branching process*, and the techniques employed above are typical of those applied in the modern theory of branching processes (chapter 8). Equation (6.2.6) should be noted. The arguments in $\phi_t((\frac{1}{2}+\frac{1}{2}z_1), z_0^2)$ are the probability-generating functions associated with individuals in the age groups 0 and 1 respectively. Many of the processes encountered in later chapters are branching processes.

6.3 *Bartlett's more general model*

Consider the time axis divided into small segments of length Δt. Using this time unit, we consider the points of time ..., -3, -2, -1, 0, 1, ... units. The following assumptions are made:

(*a*) The chance of an individual, born in time interval $(r, r+1)$ and alive at time s, surviving to time $s+1$ is $1-\mu(s-r)\,\Delta t+o(\Delta t)$; (the individual is $s-r$ units old next birthday);

(*b*) the chance of an individual, born in the interval $(r, r+1)$

and alive at time s, giving birth to a new individual in the interval $(s, s+1)$, surviving at time $s+1$ is $\lambda(s-r)\,\Delta t + o(\Delta t)$;

(*c*) the number of individuals at time s born in the interval $(r, r+1)$ is $N_r(s)$;

(*d*) multiple births and the effect of the gestation period etc. are all neglected; and

(*e*) all the probabilities at each transition act independently.

The time unit Δt is very small, and the probability of two or more vital events occurring in the population during that time is of order $(\Delta t)^2$. The following possibilities for group r during the time unit $(s, s+1)$ need to be considered.

(i) *Death and no birth.* The probability of this event is

$$N_r(s)\,\mu(s-r)\,\Delta t + o(\Delta t).$$

If the event takes place, $N_r(s+1) = N_r(s) - 1$ and $N_s(s+1) = 0$.

(ii) *Birth and no death.* The probability of this event is

$$N_r(s)\,\lambda(s-r)\,\Delta t + o(\Delta t).$$

If the event takes place, $N_r(s+1) = N_r(s)$ and $N_s(s+1) = 1$.

(iii) *No vital event.* The probability that no vital event takes place is $1 - N_r(s)\,\{\mu(s-r) + \lambda(s-r)\}\,\Delta t + o(\Delta t)$.

The moment-generating function of the population at time $s+1$ may be written in the form

$$M_{s+1}(\boldsymbol{\theta}) = \underset{\{N_r(s)\}}{\mathscr{E}}\ \underset{\{N_r(s+1)|N_r(s)\}}{\mathscr{E}} \exp\left(\sum_{r=-\infty}^{s} \theta_r N_r(s+1)\right).$$

According to comments (i), (ii) and (iii) above, the right-hand side may be expanded as follows:

$$\underset{\{N_r(s)\}}{\mathscr{E}} \exp\left(\sum_{r=-\infty}^{s-1} \theta_r N_r(s)\right)\left\{\sum_{j=-\infty}^{s-1} N_j(s)\,\mu(s-j)\,\Delta t\, \mathrm{e}^{-\theta_j} \quad \text{single death}\right.$$

$$+ \sum_{j=-\infty}^{s-1} N_j(s)\,\lambda(s-j)\,\Delta t\, \mathrm{e}^{\theta_s} \quad \text{single birth}$$

$$\left. + 1 - \sum_{j=-\infty}^{s-1} N_j(s)\,(\mu(s-j) + \lambda(s-j))\,\Delta t\right\} \quad \text{no event}$$

and after some manipulation, we find that $\Delta M_s(\boldsymbol{\theta})/\Delta t$ is equal to

$$\sum_{j=-\infty}^{s-1} \{\mu(s-j)\,(\mathrm{e}^{-\theta_j} - 1) + \lambda(s-j)\,(\mathrm{e}^{\theta_s} - 1)\}\,\frac{\partial}{\partial \theta_j} M_s(\boldsymbol{\theta}).$$

The cumulant-generating function $K_s(\boldsymbol{\theta}) = \log M_s(\boldsymbol{\theta})$ is now defined, and we note that

$$\frac{\mathrm{d}}{\mathrm{d}t}M_t(\boldsymbol{\theta}) = M_t(\boldsymbol{\theta})\frac{\mathrm{d}}{\mathrm{d}t}K_t(\boldsymbol{\theta})$$

and
$$\frac{\partial}{\partial\theta_j}M_t(\boldsymbol{\theta}) = M_t(\boldsymbol{\theta})\frac{\partial}{\partial\theta_j}K_t(\boldsymbol{\theta}).$$

To the same order of approximation therefore, $\Delta K_s(\boldsymbol{\theta})/\Delta t$ is equal to

$$\sum_{j=-\infty}^{s-1} \{\mu(s-j)(\mathrm{e}^{-\theta_j}-1)+\lambda(s-j)(\mathrm{e}^{\theta_s}-1)\}\frac{\partial}{\partial\theta_j}K_s(\boldsymbol{\theta}). \quad (6.3.1)$$

Let us now define e_r as the expected value of $N_r(s)$, $C_{r,r}$ as the variance of $N_r(s)$ and $C_{p,q}$ as the covariance of $N_p(s)$ and $N_q(s)$. The coefficients of the powers of $\{\theta_j\}$ in the expansion of the above relationship for $K(\boldsymbol{\theta})$ are equated to yield the following results:

$$\Delta e_s \cong \sum_{j=-\infty}^{s-1} \lambda(s-j)\, e_j \Delta t; \quad (6.3.2)$$

$$\Delta e_q \cong -\mu(s-q)\, e_q \Delta t; \quad (6.3.3)$$

$$\Delta C_{ss} \cong \sum_{j=-\infty}^{s-1} \lambda(s-j)\, e_j \Delta t; \quad (6.3.4)$$

$$\Delta C_{sq} \cong \sum_{j=-\infty}^{s-1} \lambda(s-j)\, C_{jq} \Delta t; \quad (6.3.5)$$

$$\Delta C_{qq} \cong \{\mu(s-q)\, e_q - 2\mu(s-q)\, C_{qq}\}\Delta t; \quad (6.3.6)$$

and
$$\Delta C_{pq} \cong -\{\mu(s-p)+\mu(s-q)\}\, C_{pq} \Delta t. \quad (6.3.7)$$

The subscripts p and q are both less than s. It is perhaps worth noting that the change in the total population is given by

$$\Delta e \cong \sum_{r=-\infty}^{s-1} \{\lambda(s-r)-\mu(s-r)\}\, e_r \Delta t, \quad (6.3.8)$$

and this result can be confirmed by general reasoning.

The results of this section should be compared with those in chapter 9. The formulation in discrete time by Bartlett was designed to give asymptotic results when the time interval was made small. In this sense, Bartlett's model may be said to lie half-way between a discrete formulation and a continuous formulation.

6.4 *The continuous-time model of D. G. Kendall*

Bartlett did not attempt to solve the equations for the general model in his notes. However, in 1949 a stochastic version of the classical (Sharpe and Lotka) continuous-time model appeared. This model, presented by D. G. Kendall at the Symposium on Stochastic Processes of the Royal Statistical Society, was inspired by Bartlett's work.

Kendall considered one sex only, and defined a function $N(x, t)$ to describe the state of the stochastic process at time t, in the sense that the Stieltjes integral

$$\int_{x_1}^{x_2} \mathrm{d}N(x,t) \tag{6.4.1}$$

is the actual number of individuals in the age group $(x_1, x_2]$ at time t. The following assumptions are made:

(*a*) The subpopulations generated by two co-existing individuals develop in complete independence of one another.

(*b*) An individual at age x existing at the epoch t has a chance $\lambda(x)\,\mathrm{d}t + o(\mathrm{d}t)$ of producing a new individual of age zero during the subsequent time interval of length $\mathrm{d}t$.

(*c*) The birth rate $\lambda(x)$ varies in any manner with the age x of the parent, but is independent of the epoch t.

(*d*) An individual of age x existing at the epoch t has a chance $\mu_x \mathrm{d}t + o(\mathrm{d}t)$ of dying during the subsequent time interval of length $\mathrm{d}t$.

(*e*) The death rate μ_x varies in any manner with the age x, but is independent of the epoch t.

The actual problems of distribution are extremely difficult, but it is possible to do something with the moments. In the discrete model, the function $\mathrm{d}N(x, t)$ is replaced by the vector variable $\mathbf{N}(t)$, whose components $\{N_r(t)\}$ enumerate the numbers of individuals in each of the several age categories. (See section 6.3.) The simplest description of the process is then in terms of the moment-generating function

$$\mathscr{E} \exp\left(\sum_r \theta_r N_r(t)\right), \tag{6.4.2}$$

from which moment recurrence relations can be derived in the manner of Bartlett. This at once suggests that the continuous model can most conveniently be described by means of the *moment-generating functional*

$$M(\theta(.);t) = \mathscr{E} \exp\left(\int_{x=0}^{\infty} \theta(x)\,\mathrm{d}N(x,t)\right), \tag{6.4.3}$$

where $\theta(x)$ is now an arbitrary function of the age x.

Following Bartlett's method, $M(\theta(.);t+dt)$ is equal to

$$\underset{\{N(x,t)\}}{\mathscr{E}}\ \underset{\{N(x,t+dt)|N(x,t)\}}{\mathscr{E}} \exp\left(\int_{x=0}^{\infty} \theta(x)\, dN(x,t+dt)\right),$$

and this expression may be expanded in the same manner as before. The notation is a little different, since Bartlett categorizes individuals according to their date of birth, whereas Kendall categorizes them according to their ages. Again, the chance of more than one event for the whole population (birth or death) is of smaller order than dt:

$$\underset{\{N(x,t)\}}{\mathscr{E}} \exp\left(\int_{y=dt}^{\infty} \theta(y)\, dN(y-dt,t)\right)\left\{\int_0^{\infty} dN(x,t)\,\mu_x\, dt\, e^{-\theta(x+dt)} \quad \text{single death}\right.$$

$$+\int_0^{\infty} dN(x,t)\,\lambda(x)\, dt\, e^{\theta(0)} \quad \text{single birth}$$

$$\left.+1-\int_0^{\infty} dN(x,t)\,(\mu_x+\lambda(x))\, dt\right\}. \quad \text{no event}$$

After some manipulation, we find that $\partial M(\theta(.);t)/\partial t$ is equal to

$$\int_0^{\infty} \{\mu_x(e^{-\theta(x)}-1)+\lambda(x)(e^{\theta(0)}-1)\}\frac{\partial}{\partial\theta(x)} M(\theta(.);t)$$
$$+\{M(\theta(.+dt);t)-M(\theta(.);t)\}/dt.$$

The *cumulant-generating functional* $K(\theta(.);t)$ is defined equal to

$$\log M(\theta(.);t),$$

and the above equation is still true when M is replaced by K. If only linear and quadratic terms in θ are required, we may equate

$$\frac{\partial}{\partial t} K(\theta(.);t)$$

to

$$\int_0^{\infty} \{-\mu_x\theta(x)+\lambda(x)\,\theta(0)+\tfrac{1}{2}\mu_x(\theta(x))^2+\tfrac{1}{2}\lambda(x)\,(\theta(0))^2\}$$
$$\times\frac{\partial}{\partial\theta(x)} K(\theta(.);t)+\{K(\theta(.+dt);t)-K(\theta(.);t)\}/dt. \quad (6.4.4)$$

But the *cumulant-generating functional* $K(\theta(.);t)$ may be expanded as follows

$$\int_0^{\infty} \theta(x)\,\alpha(x,t)\, dx+\frac{1}{2}\int_0^{\infty} \{\theta(x)\}^2\beta(x,t)\, dx$$
$$+\frac{1}{2}\int_0^{\infty}\int_0^{\infty} \theta(x)\,\theta(y)\,\gamma(x,y,t)\, dx\, dy+\ldots$$

where the cumulant functions $\alpha(x,t)$, $\beta(x,t)$ and $\gamma(x,y,t)$ are defined as follows:

$$\left.\begin{aligned} \mathscr{E}\,\mathrm{d}N(x,t) &= \alpha(x,t)\,\mathrm{d}x + o(\mathrm{d}x),\\ \operatorname{Var}\mathrm{d}N(x,t) &= \beta(x,t)\,\mathrm{d}x + o(\mathrm{d}x),\\ \operatorname{Cov}\{\mathrm{d}N(x,t),\mathrm{d}N(y,t)\} &= \gamma(x,y,t)\,\mathrm{d}x\,\mathrm{d}y + o(\mathrm{d}x\,\mathrm{d}y)\,(x \neq y). \end{aligned}\right\} \tag{6.4.5}$$

When formula (6.4.4) is expanded, and the coefficients of the powers of $\theta(x)$ are equated, the following differential equations emerge:

$$\left.\begin{aligned} \frac{\partial\alpha}{\partial t}+\frac{\partial\alpha}{\partial x} &= -\mu\alpha,\\ \frac{\partial\beta}{\partial t}+\frac{\partial\beta}{\partial x} &= \mu\alpha-2\mu\beta,\\ \text{and}\qquad \frac{\partial\gamma}{\partial t}+\frac{\partial\gamma}{\partial x}+\frac{\partial\gamma}{\partial y} &= -(\mu_x+\mu_y)\,\gamma. \end{aligned}\right\} \tag{6.4.6}$$

This was Kendall's first method of attacking the problem. Difficulties now arise because of the discontinuities in $\alpha(x,t)$, $\beta(x,t)$, and $\gamma(x,y,t)$ at the points $x = t$ and $y = t$. These discontinuity points are caused by the initial members of the population; the only individuals aged greater than t are those who were in the population at time $t = 0$, and these clearly do not have an expectation density $\alpha(x,t)$ or variance/covariance densities. An alternative simpler method of analysis is available, and this is discussed in the following section.

6.5 *An alternative analysis of Kendall's model*

In his 1949 paper, D. G. Kendall also gave an alternative simpler treatment of this stochastic model. The following lemma is required.

Lemma 6.5.1. Let M and N be discrete correlated random variables taking non-negative integral values. Let their means be μ and ν respectively, their variance σ^2 and τ^2 respectively, and their covariance $\operatorname{Cov}(M,N)$. Let M_1' be a random variable conditional on M, and having the conditional binomial distribution $B(M,p_1)$. Similarly, let M_2' and N' be random variables having the conditional binomial distributions $B(M,p_2)$ and $B(N,p)$, respectively. Then, if the conditional distributions are mutually independent,

$$\mathscr{E}M_1' = p_1\mu, \tag{6.5.1}$$
$$\operatorname{Var} M_1' = p_1^2\sigma^2 + p_1 q_1 \mu, \tag{6.5.2}$$
$$\operatorname{Cov}(M_1', M_2') = p_1 p_2 \sigma^2, \tag{6.5.3}$$
$$\operatorname{Cov}(M_1', N') = p_1 p \operatorname{Cov}(M, N), \tag{6.5.4}$$

where $q_1 = 1 - p_1$. Furthermore, if M is a binomial random variable, then M_1' is also a binomial random variable.

Proof. The proofs of these results are simple. Consider, for example the expected value of $M_1'(M_1' - 1)$.

$$\begin{aligned}
\mathscr{E}M_1'(M_1' - 1) &= \sum_{j=0}^{\infty} j(j-1) \Pr(M_1' = j) \\
&= \sum_{m=0}^{\infty} \sum_{j=0}^{m} j(j-1) \Pr(M_1' = j | M = m) \Pr(M = m) \\
&= \sum_{m=0}^{\infty} \sum_{j=0}^{m} j(j-1) \binom{m}{j} p_1^j q_1^{m-j} \Pr(M = m) \\
&= \sum_{m=0}^{\infty} p_1^2 m(m-1) \Pr(M = m) \\
&= p_1^2 \mathscr{E}M(M-1).
\end{aligned}$$

When this result is combined with the result for the expectation of M_1', equation (6.5.2) results. The other formulae are similarly proved.

Let us suppose that initially there is just one member of the population aged X, so that

$$\left.\begin{aligned} &N(x, 0) = 0, x < X, \\ \text{and} \quad &N(x, 0) = 1, x > X. \end{aligned}\right\} \tag{6.5.5}$$

The fortunes of this single ancestral individual are best described by introducing a special variable $z(t)$ which is equal to 1 if the ancestor is still alive at time t, and zero otherwise. The probability that $z(t) = 1$ is given by

$$\Pr(z(t) = 1) = {}_t p_X. \tag{6.5.6}$$

More complicated initial conditions can, of course, be represented by a superposition of solutions of this type.

Consider now the random variable $dN(x, t)$, when $x < t$. Using equation (6.5.1), we see that

$$\begin{aligned}
\mathscr{E}dN(x,t) &= {}_x p_0 \mathscr{E}dN(0, t-x) \\
&= {}_x p_0 \left[\int_0^{t-x} \lambda(y) \mathscr{E}\{dN(y, t-x)\} + \lambda(X+t-x)\mathscr{E}\{z(t-x)\}\right] dx.
\end{aligned}$$

But the expected value of $z(t)$ is ${}_tp_X$, and the expected value of $dN(x,t)$ is given by equation (6.4.5). It follows therefore that

$$\alpha(x,t) = {}_xp_0\int_0^{t-x}\lambda(y)\,\alpha(y,t-x)\,dy + {}_xp_0\lambda(X+t-x)_{t-x}p_X, \tag{6.5.7}$$

and this equation is valid for $x < t$. Putting $x = 0$, and applying equation (6.5.1) again inside the integral,

$$\alpha(0,t) = \int_0^t \lambda(y)\,{}_yp_0\alpha(0,t-y)\,dy + \lambda(X+t)_tp_X. \tag{6.5.8}$$

This is the basic integral equation (3.2.2) of Sharpe and Lotka.

Similar calculations may be performed for the variance of $dN(x,t)$, defined in equation (6.4.5). We make use of equation (6.5.2) to prove that the mean and variance of $dN(0,t-x)$ are equal, and to deduce that the mean and variance of $dN(x,t)$ are also equal. It follows that $\beta(x,t)$ obeys the same integral equation as $\alpha(x,t)$ and

$$\beta(x,t) = {}_xp_0\int_0^{t-x}\lambda(y)\,\beta(y,t-x)\,dy + {}_xp_0\lambda(X+t-x)_{t-x}p_X. \tag{6.5.9}$$

Equations (6.5.7) and (6.5.9) were derived for $x < t$. We must now consider the case $x > t$. There can be only one individual alive in this range, namely the original ancestor, and he is alive with probability ${}_tp_X$. If $\alpha(x,t)$ and $\beta(x,t)$ were defined in this range, both would be zero *except* at the point $x = X+t$, where they would both be infinite. However, it is not necessary to define $\alpha(x,t)$ and $\beta(x,t)$ for $x > t$ because

$$\mathscr{E}\,dN(X+t,t) = {}_tp_x, \tag{6.5.10}$$

and

$$\operatorname{Var} dN(X+t,t) = {}_tp_X\,{}_tq_X. \tag{6.5.11}$$

There are now only the covariances to discuss, and the relevant cumulant function $\gamma(x,y,t)$ is defined in equation (6.4.5). From the nature of the problem, γ satisfies the identities

$$\gamma(x,y,t) = \gamma(y,x,t),$$

$$\gamma(x,y,t) = 0 \quad \text{for} \quad x > t \quad \textit{and} \quad y > t;$$

it need therefore only be determined when $x < t < y$ and when $x < y < t$. If $x < t < y$, $\gamma(x,y,t) = 0$ except for $y = X+t$, when $\gamma(x,y,t)$ is infinite. In this case, it is enough to have evaluated

$$\kappa(x,t)\,dx = \operatorname{Cov}\{z(t), dN(x,t)\} \quad (x < t). \tag{6.5.12}$$

From equation (6.5.4) we note that

$$\mathrm{Cov}\,(z(t), \mathrm{d}N(x,t)) = {}_xp_{X+t-x}\,{}_xp_0\,\mathrm{Cov}\,(z(t-x), \mathrm{d}N\,(0, t-x)).$$

Equations (6.5.3), (6.5.4) and (6.5.12) are then used to deduce that

$$\kappa(x,t) = {}_xp_0\,{}_xp_{X+t-x}\Big\{\lambda(X+t-x)\;{}_{t-x}p_X\;{}_{t-x}q_X + \int_0^{t-x} \lambda(y)\,\kappa(y, t-x)\,\mathrm{d}y\Big\}, \quad (6.5.13)$$

where $x < t$.

There is now only $\gamma(x,y,t)$ to consider for $x < y < t$. We note that

$$\mathrm{Cov}\,(\mathrm{d}N(x,t), \mathrm{d}N(y,t)) = {}_xp_0\,{}_xp_{y-x}\,\mathrm{Cov}\,(\mathrm{d}N(0,t-x), \mathrm{d}N(y-x, t-x)).$$

By the repeated application of formula (6.5.4), we see that this covariance is equal to

$${}_xp_0\,{}_xp_{y-x}\Big\{\lambda(X+t-x)\,\mathrm{Cov}\,(z(t-x), \mathrm{d}N(y-x,t-x))\,\mathrm{d}x + \mathrm{d}x\int_{u=0}^{t-x} \lambda(u)\,\mathrm{Cov}\,(\mathrm{d}N(u,t-x), \mathrm{d}N(y-x,t-x))\Big\},$$

and it follows that

$$\gamma(x,y,t) = {}_xp_0\,{}_xp_{y-x}\Big\{\lambda(X+t-x)\,\kappa(y-x,t-x) + \beta(y-x,t-x)\,\lambda(y-x) + \int_{u=0}^{t-x}\lambda(u)\,\gamma(y-x,u,t-x)\,\mathrm{d}u\Big\}. \quad (6.5.14)$$

The additional term $\beta(y-x, t-x)$, due to the covariance

$$\mathrm{Cov}\,(\mathrm{d}N(y-x,t-x), \mathrm{d}N(y-x,t-x))$$

inside the integral of the previous line, should be noted.

Kendall derived these integral equations ((6.5.7), (6.5.9), (6.5.13) and (6.5.14)) for the special case in which the force of mortality μ_x is constant for all ages. The equations are difficult to solve, save in some unrealistic special cases; in the original paper, for example, the case in which both $\lambda(x)$ and μ_x were independent of age was solved.

6.6 *Exercises*

1 Expand both sides of equation (6.2.4) in powers of θ_0 and θ_1. Equate the coefficients of the various powers to obtain equation (6.2.5).

2 Work through the analysis of section 6.3 *in detail*, and derive equations (6.3.2) to (6.3.8).

3 Work through the analysis of section 6.4 *in detail*, and derive equations (6.4.6). *Hint.* This question is rather more difficult than question 2, and it should only be attempted after the other question has been solved.

4 A two-age group population of the Leslie type may be represented deterministically by the following matrix:

$$\begin{pmatrix} 0 & 3 \\ \frac{1}{2} & 0 \end{pmatrix}.$$

Determine the latent roots of this matrix.

5 The population described in question 4 is actually a stochastic population in which the probability of survival to the reproductive age is an independent chance of $\frac{1}{2}$ for each individual in age group 0 and the number of offspring is fixed at three. Derive the equations for this population corresponding to formulae (6.2.4), (6.2.5) and (6.2.6), and determine the probability of extinction for the population if at time $t = 0$ there is exactly one individual and he is in age group 0.

6 The population described in question 4 is now assumed to be a stochastic population in which the probability of survival to the reproductive age is an independent chance of $\frac{1}{2}$ for each individual in age group 0. The probability of reproduction is an independent chance of $\frac{1}{2}$ for each individual in age group 1, and each reproducing individual gives birth to exactly six offspring. Derive the equations corresponding to (6.2.4), (6.2.5) and (6.2.6), and determine the probability of extinction for the population if at time $t = 0$ there is exactly one individual and he is in age group 0.

7 Equations (6.4.6) are valid when x lies in the range $0 < x < t$. Use the method of section 5.5 to derive the general solution to the first of these three equations.

8 Equations (6.4.6) are valid when x lies in the range $0 < x < t$. Derive equations for $\alpha(0, t)$ and $\beta(0, t)$ from formula (6.4.4). Deduce that $\alpha(0, t) = \beta(0, t)$.

9 M is a random variable taking non-negative integral values. M' is a random variable conditional on M and having the conditional binomial distribution $B(M, p)$. Give an expression for the third-order falling factorial moment of M' in terms of similar quantities for M.

7
The two-sex problem

7.1 *Introduction*

In the earlier chapters, population models have been discussed which are applicable to the growth of either sex. The development of the other sex is then assumed to be consistent with the assumption of constant fertility measures and constant mortality rates for the sex under consideration. Most demographers until about 1947 were preoccupied with the female component of the population, because of its shorter reproductive age range and because illegitimate births are more readily attributable to the mother, and the fact that rather contradictory results were often obtained when these one-sex models were also applied to the male sex caused them little concern. Some were rather more critical. R. R. Kuczynski (1932) calculated the male and female net reproduction rates[1] for France in 1920–3, and found the male rate to be 1.194 and the female rate to be 0.977. The use of a one-sex model with the female component of the population would predict a continually decreasing population for France whilst the same model applied to the male component would predict a continually increasing population. Kuczynski explained the difference in the rates as being due to the effects of male war casualties.

Two attempts were made in 1947–8 to reconcile the male and female net reproduction rates. The paper by P. H. Karmel appeared first in 1947, giving a detailed discussion of the problems involved, and some numerical calculations. In the second paper, by A. H. Pollard in 1948, a two-sex population model was suggested, and it is this model that is discussed in the following section. Several other papers dealing with the problem have since appeared.

[1] The net reproduction rate R_0 was defined in section 3.4. It is equal to the average number of daughters (sons) that will be born to a female (male) now aged exactly 0. The females (males) are exactly replacing themselves when $R_0 = 1$.

7.2 *The deterministic model of A. H. Pollard*

To eliminate some of the inconsistencies inherent in the various one-sex population models and in the rates computed using the models, A. H. Pollard (1948) considered male births to females and female births to males. He defined $M(t)\,\mathrm{d}t$ to be the number of male births in time element $(t, t+\mathrm{d}t)$ and $F(t)\,\mathrm{d}t$ to be the number of female births during that time element. The mathematical techniques used are similar to those of F. R. Sharpe and A. J. Lotka. Both males and females are assumed to be non-fertile for ages less than α and ages greater than β. We obtain the following integral equations when $t > \beta$.

$$F(t) = \int_\alpha^\beta M(t-x)\,{}_x{}^m p_0 \lambda_{mf}(x)\,\mathrm{d}x; \tag{7.2.1}$$

and
$$M(t) = \int_\alpha^\beta F(t-y)\,{}_y{}^f p_0 \lambda_{fm}(y)\,\mathrm{d}y. \tag{7.2.2}$$

The functions ${}_x{}^m p_0$ and ${}_y{}^f p_0$ are the male and female survival probabilities, and $\lambda_{mf}(x)$ and $\lambda_{fm}(y)$ are the female-birth rate to males and the male-birth rate to females respectively. Equations (7.2.1) and (7.2.2) may be combined as follows:

$$F(t) = \int_\alpha^\beta\int_\alpha^\beta F(t-x-y)\,\phi(x).\xi(y)\,\mathrm{d}x\,\mathrm{d}y; \tag{7.2.3}$$

and
$$M(t) = \int_\alpha^\beta\int_\alpha^\beta M(t-x-y)\,\phi(x)\,\xi(y)\,\mathrm{d}x\,\mathrm{d}y, \tag{7.2.4}$$

where
$$\left.\begin{aligned}\phi(x) &= {}_x{}^m p_0 \lambda_{mf}(x),\\ \text{and}\quad \xi(y) &= {}_y{}^f p_0 \lambda_{fm}(y).\end{aligned}\right\} \tag{7.2.5}$$

The total birth rate at time t, $B(t) = F(t) + M(t)$ therefore satisfies the equation

$$B(t) = \int_\alpha^\beta\int_\alpha^\beta B(t-x-y)\,\phi(x)\,\xi(y)\,\mathrm{d}x\,\mathrm{d}y. \tag{7.2.6}$$

Equations (7.2.3), (7.2.4) and (7.2.6) are all of the same form, and their solutions therefore will also be of the same form. When the trial solution $B(t) = B\mathrm{e}^{st}$ is substituted in equation (7.2.6), we obtain

$$\int_\alpha^\beta\int_\alpha^\beta \mathrm{e}^{-(x+y)s}\phi(x)\,\xi(y)\,\mathrm{d}x\,\mathrm{d}y = 1. \tag{7.2.7}$$

This equation is obtained whether the solution is sought for equation (7.2.3), or (7.2.4) or (7.2.6), and hence the values of s obtained apply to male, female and total births. The proof of the following theorem follows that of theorem 3.2.1.

Theorem 7.2.1. The integral equation (7.2.7.) has exactly one real solution $s = s_0$. Any complex roots $\{s_j\}$ occur in complex conjugate pairs, and $s_0 >$ real (s_j).

Assuming[2] that all the solutions to equation (7.2.6) are of the form $B(t) = Be^{st}$, we conclude that the behaviour of the population is determined asymptotically by the real root s_0 and a stable age/sex distribution exists.

This two-sex model is very artificial, but it is important to remember the purpose of Pollard's analysis. He was seeking a measure of population increase which avoided the logical inconsistencies of Lotka's method, and the above provided such a measure. If we assume that the ratio of male to female births is constant, independent of age or sex of parent and equal to X, then it is easy to prove that s_0 lies between the intrinsic rate of increase for the male population and the intrinsic rate of increase for the female population. Furthermore, s_0 can be expressed approximately as a linear function of the male and female intrinsic rates of increase.

7.3 *The two-sex deterministic models of D. G. Kendall*

In his 1949 paper, presented in 1948 at the Royal Statistical Society Symposium on Stochastic Processes, D. G. Kendall mentioned the problem of the two sexes, and suggested a few different deterministic approaches. He first considered the simplest one-sex deterministic model characterized by the differential equation

$$F' = (\lambda - \mu)\,F, \tag{7.3.1}$$

where $F(t)$ is the number of females alive at time t, and λ and μ are the instantaneous birth and death rates respectively. The most natural way to generalize equation (7.3.1) is to write

$$\left.\begin{aligned} M' &= -\mu M + \tfrac{1}{2}\Lambda(M,F), \\ F' &= -\mu F + \tfrac{1}{2}\Lambda(M,F), \end{aligned}\right\} \tag{7.3.2}$$

where $\Lambda(M,F)$ is symmetric in M and F and represents the contribution from the birth rate. As a first approximation, it is assumed that the death rate is the same for both sexes, and that each birth is equally

[2] See sections 3.2 and 4.6.

likely to be male or female. Subtracting the second equation from the first,

$$\frac{\mathrm{d}}{\mathrm{d}t}(M-F) = -\mu(M-F),$$

so that

$$M(t)-F(t) = (M(0)-F(0))\,\mathrm{e}^{-\mu t}. \tag{7.3.3}$$

Thus, any initial excess of one sex over the other will disappear in the course of time. To proceed further with models of this type, some assumption about the form of the birth function $\Lambda(M, F)$ is necessary.

In an attempt to represent random mating, one might assume Λ to be proportional to MF; however, the number of births per unit time then varies as the square of the total population size for a constant sex ratio, and the solution is of an unstable nature. Consider, for example, the case in which $M(0) = F(0)$. From equation (7.3.3), $M(t) = F(t)$ for all t and M and F both satisfy an equation of the form

$$M' = -\mu M\left(1-\frac{M}{\alpha}\right), \tag{7.3.4}$$

where α is constant. This differential equation can be solved, and the solution obtained depends upon the initial conditions. If $M(0)$ is less than α, $M(t)$ tends to zero as t tends to infinity; if $M(0)$ is equal to α, $M(t)$ is constant for all t; and if $M(0)$ is greater than α, $M(t)$ tends to infinity as t tends to a finite positive value. In this third case, there is literally a population explosion. Similar but more complicated phenomena occur when $F(0) \neq M(0)$.

The difficulties encountered above are avoided if Λ is linear in the total population size, and in particular if Λ is set equal to $2\lambda\sqrt{(MF)}$. The basic equations (7.3.2) are easily solved in this case by writing $M = R^2$ and $F = S^2$. We find that M and F tend jointly to infinity if $\lambda > \mu$, and to zero if $\lambda < \mu$, as $t \to \infty$. When $\lambda = \mu$, $R+S$ is constant and M and F tend to the same non-zero limit.

A simpler model to analyse is obtained by assuming that Λ is equal to $\lambda(M+F)$. The birth rate depends on the arithmetic rather than the geometric mean of M and F. The solution behaves qualitatively in the same manner as in the previous example, and the critical relation between the constants is still $\mu = \lambda$. The model, however, is less realistic than the previous one. (Even when there are no females, births still take place!) L. A. Goodman (1967*c*) has generalized this type of model in discrete time for a population having an age structure.

Perhaps a more realistic model is that corresponding to

$$\Lambda = 2\lambda \min(M, F).$$

The differential equations are easily soluble, because the algebraic sign of $(M-F)$ is the same for all t; this follows from equation (7.3.3). If, initially, there is an excess of females,

$$M' = -\mu M + \lambda M \quad \text{and} \quad F' = -\mu F + \lambda M;$$
$$M = M(0)\,e^{(\lambda-\mu)t} \quad \text{and} \quad F = M - (M(0) - F(0))\,e^{-\mu t}.$$

The birth rate is determined by the number of males, and they are said to be *marriage dominant* (section 7.6).

7.4 *Some simple marriage models*

One can also discriminate between married and unmarried persons with simple deterministic models like those in section 7.3. Let M, F and C denote unmarried males, unmarried females and couples respectively. Then as a natural generalization

$$\left.\begin{aligned} M' &= -\mu_1 M + \lambda_1 C + \mu_2 C - K(M,F);\\ F' &= -\mu_2 F + \lambda_2 C + \mu_1 C - K(M,F);\\ C' &= -(\mu_1+\mu_2)\,C + K(M,F). \end{aligned}\right\} \tag{7.4.1}$$

The terms in these differential equations are obtained by considering deaths of single persons, births of single persons, deaths of married persons, and marriages of single persons. A subscript 1 is used to denote a male rate, and a subscript 2 to denote a female rate.

Kendall describes such a population in his 1949 paper. He assumes that the male and female birth and death rates are identical and that $K(M,F) = \nu \min(M,F)$. When the second equation is subtracted from the first, it is seen that

$$M - F = (M(0) - F(0))\,e^{-\mu t}, \tag{7.4.2}$$

so that as before, any initial excess of males or females disappears in the course of time.

Let us now suppose that there is initially an excess of females. In this case

$$M' = -(\mu+\nu)\,M + (\lambda+\mu)\,C, \tag{7.4.3}$$

and

$$C' = \nu M - 2\mu C. \tag{7.4.4}$$

Equation (7.4.4) gives us an expression for M in terms of C and its derivative, and hence an expression for M' in terms of C' and C''. These values are substituted into equation (7.4.3) to obtain

$$C'' + (3\mu+\nu)\,C' + \{2\mu^2 + \nu(\mu-\lambda)\}\,C = 0. \tag{7.4.5}$$

Hence, C is of the form $Ae^{p_1 t} + Be^{p_2 t}$, where p_1 and p_2 are the roots of the equation

$$p^2 + (3\mu + \nu)p + \{2\mu^2 + \nu(\mu - \lambda)\} = 0. \qquad (7.4.6)$$

It is easy to see that the roots are always real and distinct, and that exactly one of them is negative. It is the greater root which determines the character of the solution. $M(t)$ can be obtained using equation (7.4.4), and finally $F(t)$ is determined using the original differential equation for F'.

Kendall states that this type of model based on min (M, F) is 'perhaps the most realistic', and it certainly satisfies most of the conditions we would place upon a marriage model. The model does have some disadvantages, however.

1. Male and female birth and death rates are usually unequal. When they are made dissimilar in Kendall's model, the differential equations seem impossible to solve.

2. The extension of the model to a population with an age structure is rather difficult.

3. Consider a population in which $M = F = 100$. If ν is equal to 0.2, the marriage rate is 20 per unit time. The same rate will be recorded if we assume that $M = 100$ and $F = 10{,}000$. This seems a little artificial, since the 100 men will be under tremendous pressure to marry.

Let us now consider the case in which $K(M, F)$ is equal to $\nu\sqrt{(MF)}$, which is the geometric mean model. The equations (7.4.1) are far from simple to solve. Let us therefore make a trial solution

$$C = ce^{pt}. \qquad (7.4.7)$$

When this value is substituted into equations (7.4.1), we find that

$$M = me^{pt} \quad \text{and} \quad F = fe^{pt},$$

where

$$m = c(\lambda_1 - \mu_1 - p)/(p + \mu_1) \quad \text{and} \quad f = c(\lambda_2 - \mu_2 - p)/(p + \mu_2). \qquad (7.4.8)$$

The value of p is found from the characteristic equation

$$(p + \mu_1)(p + \mu_2)(p + \mu_1 + \mu_2)^2 = \nu^2(p - \lambda_1 + \mu_1)(p - \lambda_2 + \mu_2), \qquad (7.4.9)$$

a polynomial equation of degree 4.

Equation (7.4.9) has at least two real roots and possibly four. However only one root is relevant to our stable population problem. To see that this is true, we note that M, F and C must always be real and non-negative, and consequently m, f and c must also be real and non-negative. We conclude from equations (7.4.8) that p must lie to the right of $-\mu_1$ and $-\mu_2$ and to the left of $\lambda_1-\mu_1$ and $\lambda_2-\mu_2$. That is, the relevant root must lie to the right of the largest zero of the quartic curve in equation (7.4.9) and to the left of the smaller zero of the parabola.

This geometric-mean model has many desirable properties, but it is artificial in at least one respect. If the number of eligible males (females) greatly exceeds the number of eligible females (males), a K % increase in the number of single females (males) should result in a K % increase in the marriage rate. The present model does not follow this rule.

The arithmetic-mean marriage model is simpler to analyse. We assume that $K(M,F)$ is equal to $\frac{1}{2}\nu(M+F)$ and the basic differential equations can then be written in vector notation as follows:

$$\mathbf{n}' = \mathbf{A}\mathbf{n}, \tag{7.4.10}$$

where $\mathbf{n}$ is a 3-dimensional column vector with elements M, F, and C, and

$$\mathbf{A} = \begin{pmatrix} -\mu_1-\frac{1}{2}\nu & -\frac{1}{2}\nu & \mu_2+\lambda_1 \\ -\frac{1}{2}\nu & -\mu_2-\frac{1}{2}\nu & \mu_1+\lambda_2 \\ \frac{1}{2}\nu & \frac{1}{2}\nu & -(\mu_1+\mu_2) \end{pmatrix}. \tag{7.4.11}$$

This system of differential equations has an explicit solution. If the latent roots p_1, p_2 and p_3 of $\mathbf{A}$ are all distinct,[3] M, F and C are all of the form $\alpha e^{p_1t}+\beta e^{p_2t}+\gamma e^{p_3t}$. The characteristic equation for the system may be written in the form

$$(p+\mu_1)(p+\mu_2)(p+\mu_1+\mu_2)$$
$$= \tfrac{1}{4}\nu\lambda_1\lambda_2-\nu(p-\tfrac{1}{2}\lambda_1+\mu_1)(p-\tfrac{1}{2}\lambda_2+\mu_2). \tag{7.4.12}$$

The three roots of this equation may be real, in which case at least two will be negative. Alternatively, one root may be real and the other two complex.

Although this model can be analysed mathematically, it is very artificial. It is possible for the number of marriages to exceed the number of available males or females, and we then find a negative number of single males or females in the community. The model can also be criticized for the same reason as the geometric-mean model.

Having examined marriage models involving the geometric and arithmetic means of M and F, it is perhaps not unreasonable to set $K(M,F)$ equal to $2\nu MF/(M+F)$ and study the harmonic-mean model. The differential equations (7.4.1) are very difficult to solve,

[3] The complication of repeated roots is easily dealt with.

and we again make the trial solution (7.4.7). M and F have the same form as C and we find that equations (7.4.8) are still true. The value of p is found from the characteristic equation

$$(p+\mu_1+\mu_2)\{\tfrac{1}{4}\lambda_1\lambda_2-(p-\tfrac{1}{2}\lambda_1+\mu_1)(p-\tfrac{1}{2}\lambda_2+\mu_2)\}$$
$$=\nu(p-\lambda_1+\mu_1)(p-\lambda_2+\mu_2). \quad (7.4.13)$$

The criticisms levelled at the minimum, geometric-mean and arithmetic-mean models are no longer valid with this model, which seems to have most of the properties we require of a marriage model. The only drawback is an analytic one caused by the complicated form of the basic differential equations; the model is easy to use for calculation purposes, even when it is made age-specific.

TABLE 7.4.1. *Comparisons between certain marriage functions*

No. of single men	No. of single women	Time-rate of marriage according to the following marriage functions			
M	F	$0.2 \min(M, F)$	$0.2\,\dfrac{2MF}{M+F}$	$0.2\sqrt{(MF)}$	$0.2\,\dfrac{M+F}{2}$
100	100	20	20	20	20
100	120	20	21.8	21.9	22
100	150	20	24	24.4	25
100	200	20	26.7	28.3	30
100	1,000	20	36.4	63	110
100	2,500	20	38.5	100	260
100	3,600	20	38.9	120	370
100	10,000	20	39.6	200	1010
100	∞	20	40	∞	∞

Numerical comparisons between the four models are instructive, and the numbers in table 7.4.1 are worthy of some thought. It may be noted that all the models give substantially the same results when M and F are approximately equal. We know that the sex ratio of the single component of a human population is rarely as large as 2 to 1, and it may well be asked why we need concern ourselves with the more complicated models; the arithmetic-mean model is surely sufficient? A practical marriage model needs to be age-specific however; the number of single men aged 18 will be very different from the number of single women aged 30, and difficulties may be encountered if an age-specific arithmetic-mean model is used for projection purposes. A harmonic-mean model avoids these logical inconsistencies.

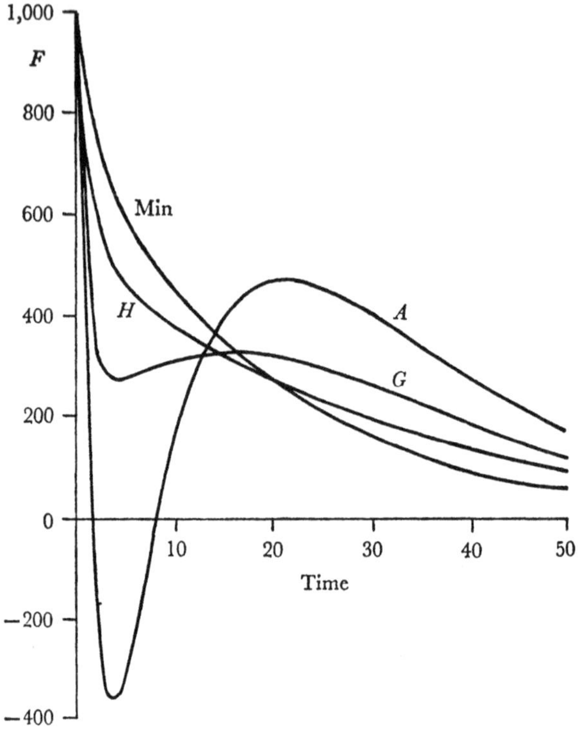

Figure 7.4.1. A numerical example. The trajectory of the female component of a population when marriages are proportional to min (M, F), $2MF/(M+F)$, $\sqrt{(MF)}$, and $\frac{1}{2}(M+F)$. The curves are labelled Min, H, G and A respectively.

A further comparison of the four models is given in figure 7.4.1, which depicts the trajectories of the female component of a population under the four different models. The curves are labelled Min, H, G and A. The initial population is composed of 10,000 males, 1,000 females and 500 couples, and the same parameter values are used in each case: $\lambda_1 = 0.115$; $\lambda_2 = 0.105$; $\mu_1 = 0.110$; $\mu_2 = 0.100$ and $\nu = 0.150$. The use of the arithmetic mean model is clearly invalid for this extreme example, and the geometric-mean model also seems unsatisfactory. Both the minimum and the harmonic-mean models seem reasonable.

7.5 *A simple model with a behavioural basis*

All the models in sections 7.3 and 7.4 were constructed by choosing a mathematical function with certain properties and using it as the marriage function. Except for the unstable product model, no real attempt was made to base the models upon human behavioural patterns.

Let us now consider a large population containing $M+F+2C$ individuals. Because of the size of the population, an individual cannot know every other individual and we assume that he (or she) knows W other persons. If encounters in the population are purely random a proportion $M/(M+F+2C)$ of the acquaintances of a single female will be single men. Assume that the probability that a particular single male/single female relationship at time t will result in marriage during time element $(t, t+\mathrm{d}t)$ is $\theta\,\mathrm{d}t$. Then the number of marriages in the population during the time element $\mathrm{d}t$ must be $F\{WM/(M+F+2C)\}\theta\,\mathrm{d}t$. That is, we choose as our marriage function

$$K(M,F) = \nu MF/(M+F+2C). \tag{7.5.1}$$

This marriage model has the same desirable properties as its close relative the harmonic-mean model. The differential equations are difficult to solve, and we again make the trial solution (7.4.7). Equations (7.4.8) are true, and the characteristic equation is

$$(\lambda_1+\lambda_2)\,(p+\mu_1+\mu_2)\left(p+\frac{\lambda_1\mu_2+\lambda_2\mu_1}{\lambda_1+\lambda_2}\right) = \nu(p-\lambda_1+\mu_1)\,(p-\lambda_2+\mu_2). \tag{7.5.2}$$

The root p relevant to the stable population must lie to the right of $-\mu_1$ and $-\mu_2$ and to the left of $\lambda_1-\mu_1$ and $\lambda_2-\mu_2$. This result follows again from equations (7.4.8), and it is true for all the models in section 7.4.

7.6 *An early two-sex model due to L. A. Goodman*

Kendall's two-sex models aroused the interest of L. A. Goodman, and the latter published some generalizations of Kendall's work in 1953. Many of the models analysed were deterministic, and in all of them the problem of age was ignored. Goodman did discuss a few stochastic models, and it is one of them we now consider.

Let us imagine that a population behaves in accordance with the following rules:

(*a*) the sub-populations generated by two co-existing individuals develop in complete independence of one another;

(*b*) a female alive at time t has a chance $\lambda_1\mathrm{d}t+o(\mathrm{d}t)$ of giving birth to a male and a chance $\lambda_2\mathrm{d}t+o(\mathrm{d}t)$ of giving birth to a female during the following time interval of length $\mathrm{d}t$;

(*c*) a female alive at time t has a chance $\mu_2 \mathrm{d}t + o(\mathrm{d}t)$ of dying and a male alive at time t a chance $\mu_1 \mathrm{d}t + o(\mathrm{d}t)$ of dying during the following time interval of length dt.

If we define $P_{m,n} \equiv P_{m,n}(t)$ as the probability that there are m men and n women in the population at time t, we can use the birth and death process technique of section 5.2 to derive the following differential-difference equation:

$$P_{m,n}' = (m+1)\,\mu_1 P_{m+1,n} + (n+1)\,\mu_2 P_{m,n+1} + n\lambda_1 P_{m-1,n} + (n-1)\,\lambda_2 P_{m,n-1} - \{m\mu_1 + n\mu_2 + n\lambda_1 + n\lambda_2\}\,P_{m,n}. \quad (7.6.1)$$

Differential equations for the moments of $M(t)$ and $F(t)$ (the number of males and females at time t) can be found by multiplying equation (7.6.1) by the appropriate powers of m and n and summing for all possible values of m and n. The technique was demonstrated in section 5.4. For the means $\bar{M} \equiv \bar{M}(t)$ and $\bar{F} \equiv \bar{F}(t)$, the following equations are obtained:

$$\bar{M}' = -\mu_1 \bar{M} + \lambda_1 \bar{F} \quad \text{and} \quad \bar{F}' = -\mu_2 \bar{F} + \lambda_2 \bar{F}. \quad (7.6.2)$$

These are the obvious deterministic equations for this type of model.

In this population, all births are attributed to the females, who are said to be *marriage dominant*.[4] It is soon apparent that the female component of the population follows the rules of a linear birth and death process, and the results of section 5.4 can be used. We know therefore that the mean and variance of the female component of the population will be given by formulae (5.4.5) and (5.4.7) with λ_2 and μ_2 substituted for λ and μ respectively. The algebra involved in determining the first- and second-order moments of $M(t)$ and the covariance Cov $(M(t), F(t))$ is somewhat tedious, although the principles involved are not difficult. The results are of little interest in the present context.

The same type of model can be used when the males are marriage dominant, and a model corresponding to a mixture of male dominance and female dominance is not difficult to construct. It should be noted that a model in which one sex has complete marriage dominance is essentially a one-sex model.

It is possible to construct a model in which female births are attributable to the males, and male births are attributable to the females. The analysis would be similar to that given above and the model would be a non-age-specific stochastic analogue of the deterministic model described in section 7.2.

[4] The use of the term 'marriage dominance' with this precise mathematical meaning appears to have originated with Goodman, although P. H. Karmel did use this expression.

7.7 *Age-specific dominance models*

The concept of female marriage dominance was introduced in section 7.6. The model described in that section ignored the problem of age. However, an obvious discrete-time extension of the model can readily allow for this factor.

The males and females are considered at discrete points of time $t = 0, 1, 2, \ldots$, and are categorized according to age and sex. Let there be $k+1$ age groups for males and $k'+1$ age groups for females. The female age groups are listed below the male age groups, and we consider a column vector of length $k+k'+2$. The numbers of males and females in age group x at time t are denoted by $n_{x,t}$ and $n_{x,t}'$ respectively. The male survivorship proportions are $\{P_x\}$ and the female survivorship proportions are $\{P_x'\}$. The case of female marriage dominance will be considered, so that all the births are attributable to the mothers. The birth rate of males to females aged x will be denoted by F_x, and the birth rate of females to females aged x by F_x'. Then we have equation (7.7.1).

$$\begin{pmatrix} n_{0,t+1} \\ n_{1,t+1} \\ n_{2,t+1} \\ \vdots \\ n_{k,t+1} \\ \hline n_{0,t+1}' \\ n_{1,t+1}' \\ n_{2,t+1}' \\ \vdots \\ n_{k',t+1}' \end{pmatrix} = \left(\begin{array}{ccccc|cccc} 0 & & & & & F_0 & F_1 & \cdots & F_{k'} \\ P_0 & & & & & & & & \\ & P_1 & & & & & & & \\ & & \ddots & & & & & & \\ & & & P_{k-1} & 0 & & & & \\ \hline & & & & & F_0' & F_1' & \cdots & F_{k'}' \\ & & & & & P_0' & & & \\ & & & & & & P_1' & & \\ & & & & & & & \ddots & \\ & & & & & & & P_{k'-1}' & 0 \end{array}\right) \begin{pmatrix} n_{0,t} \\ n_{1,t} \\ n_{2,t} \\ \vdots \\ n_{k,t} \\ \hline n_{0,t}' \\ n_{1,t}' \\ n_{2,t}' \\ \vdots \\ n_{k',t}' \end{pmatrix}. \tag{7.7.1}$$

The case of male dominance can be dealt with in the same manner. Both these cases are really only one-sex models however. (The non-zero latent roots of the recurrence matrix in equation (7.7.1) for example are those of the lower right-hand sub-matrix, which is the Leslie matrix for the female component of the population.) In practice, the number of births depends upon the number of males and the number of females. An intermediary degree of dominance would seem a reasonable choice of model, in which case some of the male births are attributed to the males and some to the females, and some of the female births are attributed to the females and some to the males. Both the on-diagonal blocks of the recurrence matrix will then have the Leslie form, and both the off-diagonal blocks will be zero except in their first rows. Such a model may avoid many of the inconsistencies mentioned in section 7.1, but there is nothing to suggest that the model is any closer to the truth than a one-sex model. Indeed, the intermediary-dominance model will still allow births to occur even when there are no females in the population! With these models it is possible, however, to prove certain rather elegant mathematical results concerning the age-sex composition of the population. The interested reader should refer to a paper by L. A. Goodman (1967*c*).

7.8 *An example*

In this example, we develop the discrete-time analogue of the two-sex model in section 7.2. The characteristic equation of the matrix involved is derived using the recurrence equation method of section 4.6. Finally, an inequality between the dominant latent roots of the two-sex matrix and the two related one-sex matrices is obtained.

The symbols k, k', $n_{x,t}$, $n_{x,t}'$, P_x and P_x' will be defined in the same manner as they were in section 7.7. The quantities F_x and F_x' will be defined slightly differently: F_x will be the fertility measure of female births to males and F_x' will be the fei tility measure of male births to females. We assume that the female vector is again listed below the male vector.

All the elements in the upper left-hand block of the recurrence matrix will be zero except those immediately below the diagonal and these elements take the values $\{P_j\}$ $(j = 0, \ldots, k)$. The lower right-hand block has a similar form with the non-zero elements $\{P_j'\}$ $(j = 0, \ldots, k')$. All the elements of the lower left-hand block are zero except those in the first row and these elements take the values $\{F_j\}$, $(j = 0, \ldots, k)$. The upper right-hand block has a similar form with the values

$$F_j', (j = 0, \ldots, k').$$

Consider the $n_{0,t}$ male births at time t, and the births of their mothers. We find that the following recurrence equation is true.

$$n_{0,t} = \sum_{x=0}^{k'} \sum_{y=0}^{k} (P_0' \ldots P_{x-1}' F_x') (P_0 \ldots P_{y-1} F_y) \, n_{0,t-x-y-2}.$$

This is a homogeneous recurrence equation of order $k+k'+2$, and we make a trial solution $n_{0,t} = c\lambda^t$ to discover that

$$\lambda^{k+k'+2} = \sum_{x=0}^{k'} \sum_{y=0}^{k} (P_0' \ldots P_{x-1}' F_x') (P_0 \ldots P_{y-1} F_y) \times \lambda^{k+k'-x-y}. \quad (7.8.1)$$

Let us now examine the relationship between the dominant latent root of the two-sex matrix and the dominant latent roots of the two one-sex matrices. (The conditions for positive regularity will usually be satisfied.) We shall make the assumption that the *sex ratio at birth*[5] is constant over time, independent of age and sex of parent and equal to X. It follows that the characteristic equation for the male one-sex model is

$$\lambda^{k+1} = \sum_{y=0}^{k} (P_0 P_1 \dots P_{y-1} X F_y)\, \lambda^{k-y}, \tag{7.8.2}$$

and the female characteristic equation is

$$\lambda^{k'+1} = \sum_{x=0}^{k'} (P_0' P_1' \dots P_{x-1}' X^{-1} F_x')\, \lambda^{k'-x}. \tag{7.8.3}$$

Equation (7.8.1) is equal to the product of equations (7.8.2) and (7.8.3). All three equations may be written in the form (4.4.2), and it is easy to deduce that the dominant latent root for the two-sex model must lie between the dominant latent roots for the two one-sex models.

7.9 *Exercises*

1 The two-sex analysis of A. H. Pollard is to be applied to a human population, and the joint intrinsic rate of increase s_0 of the population is to be computed. It is known that the sex ratio at birth is constant over time, independent of age and sex of parent and equal to X. The one-sex analysis of F. R. Sharpe and A. J. Lotka will also be applied to the two sexes separately, and the male and female intrinsic rates of increase will be calculated. Prove that the joint intrinsic rate of increase must lie between the two one-sex intrinsic rates.

2 In a certain population, the sex ratio at birth is constant and equal to X. The forces of mortality for males and females are constants independent of age and equal to μ_m and μ_f respectively. The birth functions $\lambda_{mf}(x)$ and $\lambda_{fm}(y)$ are independent of age and equal to λ_{mf} and λ_{fm} respectively. Determine the joint rate of natural increase s_0 of the population. Give formulae for the intrinsic rates of increase of the male and female populations considered separately.

3 A population like that described in question 2 is such that $\mu_m = \mu_f = \mu$. Prove that

$$(s_0+\mu)^2 = \{r_0\,(\text{males})+\mu\}\,\{r_0\,(\text{females})+\mu\}.$$

4 For a certain population, the sex ratio at birth is a constant X independent of age and sex. Furthermore

$${}_x^m p_0\, \lambda_{mf}(x) = M_0 c^k x^{k-1}\, e^{-cx}/\Gamma(k) \quad \text{and} \quad {}_y^f p_0 \lambda_{fm}(y) = F_0 c^j x^{j-1}\, e^{-cy}/\Gamma(j).$$

Determine the real root and the complex roots for the joint analysis of section 7.2 and prove that

$$(s_0+c)^{k+j} = \{r_0(\text{males})+c\}^k\,\{r_0(\text{females})+c\}^j.$$

[5] The *sex ratio at birth* is the ratio of male births to female births. The sex ratio at birth for Australia in 1968 was 1.0541.

5 A. H. Pollard (1948) gives the following approximate formula for s_0 in terms of r_0(males) and r_0(females)

$$s_0 \doteqdot \frac{\kappa_1(\text{males})\, r_0(\text{males}) + \kappa_1(\text{females})\, r_0(\text{females})}{\kappa_1(\text{males}) + \kappa_1(\text{females})}$$

κ_1(males) refers to the first cumulant of the net paternity function $\phi(x)$ when it is suitably scaled, and κ_1(females) refers to the first cumulant of the net maternity function $\xi(y)$ suitably scaled. Derive this formula, assuming:

(i) that the sex ratio at birth is constant and independent of age and sex of parent; and

(ii) that the functions $\phi(x)$ and $\xi(y)$ have the Gaussian form suggested by A. J. Lotka (section 3.4).

6 In a certain population, the number of births is determined in a proportion $K/(K+L)$ of generations by the male component of the population and in a proportion $L/(K+L)$ of generations by the female component of the population. The sex ratio at birth X is constant over time and independent of the age and sex of the parent concerned.

Develop a theory of population growth for this population using the techniques of section 7.2 (L. Yntema, 1952).

7 Prove the result of question 1 for the population described in question 6.

8 For the population described in question 6, prove a result analogous to that given in question 5.

9 Calculate the stable instantaneous growth rate for the arithmetic-mean, geometric-mean and harmonic-mean marriage models in the final paragraph of section 7.4.

10 Calculate the stable populations for the models in question 9.

11 Kendall's two-sex product model is described in section 7.3. When $M(0)$ is equal to $F(0)$ and they are both greater than the quantity α defined in equation (7.3.4), the population size becomes infinite for a finite value of t. What is this value of t?

12 Derive equations (7.4.8) and (7.4.9).

8
The extinction of surnames

8.1 *Introduction*

'The decay of the families of men who occupied conspicuous positions in past times has been a subject of frequent remark, and has given rise to various conjectures. It is not only the families of men of genius or those of the aristocracy who tend to perish, but it is those of all with whom history deals, in any way, even of such men as the burgesses of towns, concerning whom Mr. Doubleday has inquired and written. The instances are very numerous in which surnames that were once common have since become scarce or have wholly disappeared. The tendency is universal, and, in explanation of it, the conclusion has been hastily drawn that a rise in physical comfort and intellectual capacity is necessarily accompanied by diminution in "fertility"—using that phrase in its widest sense and reckoning abstinence from marriage as sterility. If that conclusion be true, our population is chiefly maintained through the "proletariat", and thus a large element of degradation is inseparably connected with those other elements which tend to ameliorate the race. On the other hand, M. Alphonse De Candolle has directed attention to the fact that, by the ordinary law of chances, a large proportion of families are continually dying out, and it evidently follows that, until we know what the proportion is, we cannot estimate whether any observed diminution of surnames among the families whose history we can trace, is or is not a sign of their diminished "fertility".'

In these 'Prefatory Remarks' to a paper by the Reverend H. W. Watson (1874), Francis Galton includes a quotation from A. De Candolle's work, and he then continues:

> 'although I have not hitherto published anything on the matter, I took considerable pains some years ago to obtain numerical results in respect to this very problem. I made certain very simple, but not very inaccurate, suppositions, concerning average fertility, and I worked to the nearest integer, starting with 10,000 persons, but the computation became intolerably tedious after a few steps, and I had to abandon it. More recently, having first privately applied in vain to some mathematicians, I put the problem into a shape suited to mathematical treatment, and proposed it in the pages of a well-known mathematical periodical of a high class, the "Educational Times". It met with poor success at first, because the answer it received was from a correspondent who wholly failed to perceive its intricacy, and his results were totally erroneous.'

The form in which the problem is stated in the *Educational Times* (1873) is as follows:

> 'Problem 4001: A large nation, of whom we will only concern ourselves with the adult males, N in number, and who each bear separate surnames, colonise a district. Their law of population is such that, in each generation, a_0 per cent of the adult males have no male children who reach adult life; a_1 have one such male child; a_2 have two; and so on up to a_5 who have five.
>
> Find (1) what proportion of the surnames will have become extinct after r generations; and (2) how many instances there will be of the same surname being held by m persons.'

As the only solution received was totally erroneous, Galton sought the assistance of the Reverend Henry William Watson, clergyman, mathematician and alpinist. The latter solved Galton's problem using the device now known as a probability-generating function, but due to a mathematical oversight he concluded incorrectly that all families become extinct.

Galton and Watson are usually credited with founding branching process theory (e.g. T. E. Harris, 1963; D. G. Kendall, 1966), but a correct statement of the criticality theory was given earlier by I. J. Bienaymé in 1845. This reference seems to have been completely overlooked in the branching process literature until C. C. Heyde and E. Seneta published their historical note on Bienaymé's work in 1972. The title and the opening paragraph of Bienaymé's paper show that his motivation was similar to that of Galton:

> 'On s'est beaucoup occupé de la multiplication possible du nombre des hommes; et récemment diverses observations très curieuses ont été publiées sur la fatalité qui s'attacherait aux corps de noblesse, de bourgeoisie, aux familles des hommes illustres, etc.; fatalité qui, dit-on, ferait disparaître inévitablement ce qu'on a nommé des familles fermées.'

8.2 *The Galton–Watson process*[1]

Consider a male aged 0, and let us assume that he will have exactly r sons with probability p_r $(r = 0, 1, 2, \ldots)$. The number of sons born to this man will be assumed to be independent of the population structure, and all the males in the population have sons according to the same probability law. It is useful to define a probability-generating function

$$f(z) = \sum_{r=0}^{\infty} p_r z^r \quad (|z| \leqslant 1), \tag{8.2.1}$$

where z is a complex variable. The number of males in the nth generation of the family will be denoted by Z_n, and we shall assume that $Z_0 = 1$.

According to the definition of $f(z)$, the probability of there being r males in the first generation is given by the coefficient of z^r in the expansion of $f(z)$. For the second generation, it is soon apparent that

$$\Pr(Z_2 = r|Z_0 = 1) = \sum_{x=0}^{\infty} \Pr(Z_2 = r|Z_1 = x) \times \Pr(Z_1 = x|Z_0 = 1). \tag{8.2.2}$$

It is convenient to define a probability-generating function $f_n(z)$ for the nth generation:

$$f_n(z) = \sum_{r=0}^{\infty} \Pr(Z_n = r|Z_0 = 1)\, z^r, \tag{8.2.3}$$

and using equation (8.2.2),

$$\begin{aligned} f_2(z) &= \sum_{r=0}^{\infty}\sum_{x=0}^{\infty} z^r \Pr(Z_2 = r|Z_1 = x)\Pr(Z_1 = x|Z_0 = 1) \\ &= \sum_{x=0}^{\infty} \{f(z)\}^x \Pr(Z_1 = x|Z_0 = 1) \\ &= f_1(f(z)). \end{aligned}$$

[1] Perhaps we should call it the Bienaymé process.

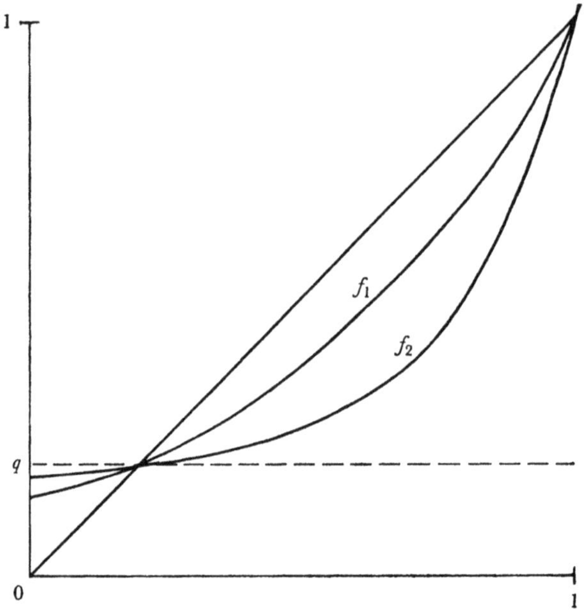

Figure 8.2.1. The graphs of $f_1(x)$ and $f_2(x)$ for the case in which the mean m is greater than one.

More generally,

$$f_{n+1}(z) = f_n(f(z)) = f(f_n(z)) \quad (n = 1, 2, \ldots), \tag{8.2.4}$$

and of course

$$f_1(z) = f(z). \tag{8.2.5}$$

Thus the function $f_n(z)$ is the nth iterate of the probability-generating function $f(z)$.

To avoid uninteresting trivialities, we shall assume that none of the probabilities $\{p_r\}$ is equal to one and that

$$p_0 + p_1 < 1. \tag{8.2.6}$$

The function $f(z)$ is then strictly convex on the unit interval (0, 1) of the real axis, and this situation is depicted in figure 8.2.1.

The first- and second-order moments of Z_n are not difficult to determine. We shall assume that the mean and variance of Z_1 are finite, and make use of the following notation:

$$m = \mathscr{E}Z_1 = \sum_{r=0}^{\infty} rp_r; \tag{8.2.7}$$

and $$\sigma^2 = \operatorname{Var} Z_1 = \sum_{r=0}^{\infty} r^2 p_r - m^2. \tag{8.2.8}$$

Let us differentiate equation (8.2.4) with respect to z, and then set z equal to one:

$$f_{n+1}'(1) = f'(f_n(1))f_n'(1) = f'(1)f_n'(1).$$

But $f_n'(1)$ is the mean of Z_n and $f'(1)$ is equal to m. Therefore, by induction

$$\mathscr{E} Z_n = m^n. \tag{8.2.9}$$

If we differentiate equation (8.2.4) twice with respect to z, and then set z equal to one, we find that

$$f_{n+1}''(1) = (m^2 + \sigma^2 - m)\, m^{2n} + m f_n''(1). \tag{8.2.10}$$

When $m = 1$, this equation may be written in the form

$$f_{n+1}''(1) - (n+1)\,\sigma^2 = f_n''(1) - n\sigma^2,$$

and we can deduce that

$$\operatorname{Var} Z_n = n\sigma^2 \quad (m = 1). \tag{8.2.11}$$

When $m \neq 1$, equation (8.2.10) may be written in the form

$$\left\{\frac{f_{n+1}''(1)}{m^{2(n+1)}} - \frac{m^2+\sigma^2-m}{m^2-m}\right\} = \frac{1}{m}\left\{\frac{f_n''(1)}{m^{2n}} - \frac{m^2+\sigma^2-m}{m^2-m}\right\}.$$

This recurrence equation gives us the second-order falling factorial moment $f_n''(1)$, and we deduce that

$$\operatorname{Var} Z_n = \frac{\sigma^2 m^n (m^n - 1)}{m^2 - m} \quad (m \neq 1). \tag{8.2.12}$$

Francis Galton was concerned with the probability q of ultimate extinction of the family surname. Clearly

$$q = \lim_{n\to\infty} \Pr(Z_n = 0) = \lim_{n\to\infty} f_n(0). \tag{8.2.13}$$

Theorem 8.2.1. If the mean number of sons born to a male

$$\mathscr{E}(Z_1 | Z_0 = 1)$$

is less than or equal to one, the probability q of extinction for the family surname is one. If the mean number is greater than one, the extinction probability is the unique non-negative solution less than one of the equation

$$z = f(z). \tag{8.2.14}$$

Proof. It is soon apparent that if q is the probability of ultimate extinction, q must be equal to $f(q)$. We therefore examine the roots of equation (8.2.14).

Because the mean and variance of Z_1 are both finite, the first two left-hand derivatives of $f(z)$ exist at $z = 1$, and they are both finite. Let us examine $f(1-h)$ where h is a positive quantity less than one. According to Taylor's theorem,

$$f(1-h) = f(1) - hf'(1) + \tfrac{1}{2}h^2 f''(1-\theta h). \qquad (8.2.15)$$

Consider first the case in which the mean number of sons is less than or equal to one. The second derivative $f''(1-\theta h)$ must be positive and finite, the first derivative $f'(1)$ must be positive and less than or equal to one, and $f(1)$ is equal to one. Therefore

$$f(1-h) > 1-h \quad (1 > h > 0).$$

That is, the curve $f(z)$ lies above the line z for $0 \leqslant z < 1$, and the only point of intersection in the unit interval is the point $z = 1$. We conclude that the surname becomes extinct with probability one.

Consider now the case in which the mean number of sons is strictly greater than one. For small h, the $f(z)$ curve will lie below the line z, and if $p_0 > 0$, the $f(z)$ curve will lie above the line z when h is closer to one. The $f(z)$ curve is strictly convex and so there exists exactly one solution to equation (8.2.14) in the range $0 < z < 1$. It would appear at first that there are two possible roots to this equation. However, the root one is rejected for the following reason: imagine that $f_n(0)$ is equal to $1-h$ where h is fairly small. According to equations (8.2.4) and (8.2.15) $f_{n+1}(0) = f(f_n(0))$ must be less than $f_n(0)$. But we know that $f_n(0)$ is a monotonic increasing function of n, and therefore it must have a limit strictly less than one. If p_0 is equal to zero, the surname cannot become extinct. This completes the proof of the theorem.

8.3 *The calculations of A. J. Lotka*

In 1931, A. J. Lotka used the methods of section 8.2 to determine the probability of extinction of the male line of descent for United States White Males 1920. He determined the following probabilities:

$p_0 = 0.4982$; $p_1 = 0.2103$; $p_2 = 0.1270$; $p_3 = 0.0730$;

$p_4 = 0.0418$; $p_5 = 0.0241$; $p_6 = 0.0132$; $p_7 = 0.0069$;

$p_8 = 0.0035$; $p_9 = 0.0015$; $p_{10} = 0.0005$; $p_n = 0.0000 \quad (n > 10)$.

According to the theory of the preceding section, we must seek a root lying between zero and one for the following equation:

$$\begin{aligned} x = 0.4982 &+ 0.2103\,x + 0.1270\,x^2 + 0.0730\,x^3 \\ &+ 0.0418\,x^4 + 0.0241\,x^5 + 0.0132\,x^6 \\ &+ 0.0069\,x^7 + 0.0035\,x^8 + 0.0015\,x^9 \\ &+ 0.0005\,x^{10}. \end{aligned} \tag{8.3.1}$$

It is not difficult to see that this population is increasing, and we know therefore that equation (8.3.1) has a non-negative root strictly less than one. It is also apparent that the root is non-zero. Writing $1-y$ for x, re-arranging and dividing by y,

$$\begin{aligned} y = \quad & 0.10700 \quad + 1.00996\,y^2 - 0.84180\,y^3 \\ & + 0.56021\,y^4 - 0.28815\,y^5 + 0.10987\,y^6 \\ & - 0.02915\,y^7 + 0.00480\,y^8 - 0.00037\,y^9. \end{aligned} \tag{8.3.2}$$

A first approximation to the root required may be obtained by ignoring powers of y above the second and solving the quadratic

$$1.00996\,y^2 - y + 0.10700 = 0. \tag{8.3.3}$$

The first approximation is found to be $y = 0.122$ and when it is substituted into the right-hand side of equation (8.3.2) we obtain the second approximation: $y = 0.12061$. Third and fourth approximations may be also obtained via equation (8.3.2). They are $y = 0.12034$ and $y = 0.12027$ respectively. It is clear therefore that the root required is $y = 0.1203$ correct to 4 decimal places. We conclude that the probability of extinction is 0.8797 correct to four places of decimals.

8.4 *The multi-type process*

The Galton–Watson process described in section 8.2 may be generalized by considering a process which involves a finite number of different types of individual. This multi-type process will occupy a central position in the theory of chapter 9.

Consider an individual of type j ($j = 1, 2, \ldots, k$), and let us assume that during a unit time interval this individual gives birth to r_1 individuals of type 1, r_2 individuals of type $2, \ldots, r_k$ individuals of type k, with probability $p_j(r_1, \ldots, r_k)$. All the individuals in the population act independently of one another, and for every j

$$\sum_{r_1} \sum_{r_2} \cdots \sum_{r_k} p_j(r_1, \ldots, r_k) = 1.$$

(With this definition of $p_j(r_1, \ldots, r_k)$, it is assumed that the individual of type j 'dies' during the unit time interval. If this is not true, the individual's existence at the following point of time may be accounted for by a suitable 'birth'.)

It is useful to define the following probability-generating function

$$f^j(z_1, \ldots, z_k) = \sum_{r_1=0}^{\infty} \ldots \sum_{r_k=0}^{\infty} p_j(r_1, \ldots, r_k)\, z_1^{r_1} \ldots z_k^{r_k}, \quad (8.4.1)$$

where $j = 1, 2, \ldots, k$, and the $\{z_i\}$ $(i = 1, 2, \ldots, k)$ are complex variables. The number of individuals of type j at time n will be denoted by $Z_{j,n}$. Let us assume that at time $t = 0$ there is exactly one individual in the population and that this single individual is of type j. According to the definition of f^j, the probability of there being r_1 individuals of type 1, r_2 of type 2, ..., r_k of type k at time $t = 1$ is given by the coefficient of $z_1^{r_1} z_2^{r_2} \ldots z_k^{r_k}$ in the expansion of f^j.

To simplify the notation in the remainder of this section, we shall follow T. E. Harris (1963) and define vectors $\{\mathbf{e}_i\}$ $(i = 1, 2, \ldots, k)$. All the elements of $\mathbf{e}_i$ except the ith are zero, and the ith element is one.

For the second period of time, it is soon apparent that

$$\Pr(\mathbf{Z}_2 = \mathbf{r} | \mathbf{Z}_0 = \mathbf{e}_j) = \sum_{\mathbf{x}} \Pr(\mathbf{Z}_2 = \mathbf{r} | \mathbf{Z}_1 = \mathbf{x}) \times \Pr(\mathbf{Z}_1 = \mathbf{x} | \mathbf{Z}_0 = \mathbf{e}_j), \quad (8.4.2)$$

where the vector $\mathbf{Z}_n$ has as its elements the $\{Z_{i,n}\}$, $(i = 1, 2, \ldots, k)$. It is convenient to define a probability-generating function

$$f_n^j(z_1, z_2, \ldots, z_k)$$

for the nth point of time:

$$f_n^j(\mathbf{z}) = \sum_{\mathbf{r}} \Pr(\mathbf{Z}_n = \mathbf{r} | \mathbf{Z}_0 = \mathbf{e}_j)\, z_1^{r_1} \ldots z_k^{r_k}. \quad (8.4.3)$$

Then, using equation (8.4.2) ,we have

$$\begin{aligned} f_2^j(\mathbf{z}) &= \sum_{\mathbf{r}} \sum_{\mathbf{x}} z_1^{r_1} \ldots z_k^{r_k} \Pr(\mathbf{Z}_2 = \mathbf{r} | \mathbf{Z}_1 = \mathbf{x}) \Pr(\mathbf{Z}_1 = \mathbf{x} | \mathbf{Z}_0 = \mathbf{e}_j) \\ &= \sum_{\mathbf{x}} \{f^1(\mathbf{z})\}^{x_1} \ldots \{f^k(\mathbf{z})\}^{x_k} \Pr(\mathbf{Z}_1 = \mathbf{x} | \mathbf{Z}_0 = \mathbf{e}_j) \\ &= f_1^j(f^1(\mathbf{z}), f^2(\mathbf{z}), \ldots, f^k(\mathbf{z})). \end{aligned}$$

More generally,

$$\left.\begin{aligned} & f_{n+1}^j(\mathbf{z}) = f_n^j(\mathbf{f}(\mathbf{z})) = f^j(\mathbf{f}_n(\mathbf{z})), \\ \text{or} \quad & \mathbf{f}_{n+1}(\mathbf{z}) = \mathbf{f}_n(\mathbf{f}(\mathbf{z})) = \mathbf{f}(\mathbf{f}_n(\mathbf{z})), \end{aligned}\right\} \quad (8.4.4)$$

and of course

$$\mathbf{f}_1(\mathbf{z}) = \mathbf{f}(\mathbf{z}). \tag{8.4.5}$$

Clearly

$$\mathbf{f}_{n+N}(\mathbf{z}) = \mathbf{f}_N(\mathbf{f}_n(\mathbf{z})). \tag{8.4.6}$$

This is the basic iterative functional relationship. It should be noted that the first five equations of this section have a one-to-one relationship with the first five equations of section 8.2.

We shall define an expectation matrix $\mathbf{M} = (m_{ij})$ as follows:

$$m_{ij} = \mathscr{E}(z_{i,1}|\mathbf{Z}_0 = \mathbf{e}_j) = \frac{\partial}{\partial z_i} f^j(1, 1, \ldots, 1). \tag{8.4.7}$$

To obtain a relationship for the expectation behaviour of the population, let us differentiate the first part of equation (8.4.4) partially with respect to z_i:

$$\frac{\partial}{\partial z_i} f_{n+1}{}^j(\mathbf{z}) = \sum_l \frac{\partial}{\partial f^l}\{f_n{}^j(\mathbf{f}(\mathbf{z}))\}\frac{\partial f^l}{\partial z_i}. \tag{8.4.8}$$

If we now set all the $\{z_l\}$ equal to one, we obtain

$$\mathscr{E}(Z_{i,n+1}|\mathbf{Z}_0 = \mathbf{e}_j) = \sum_l m_{il}\mathscr{E}(Z_{l,n}|\mathbf{Z}_0 = \mathbf{e}_j)$$

and conclude that

$$\mathscr{E}(\mathbf{Z}_{n+N}|\mathbf{Z}_n) = \mathbf{M}^N\mathbf{Z}_n. \tag{8.4.9}$$

This result[2] should be compared with the Leslie equation (4.2.4).

To derive a relationship for the quadratic moments, let us differentiate equation (8.4.8) partially with respect to z_s:

$$\frac{\partial^2}{\partial z_i \partial z_s} f_{n+1}{}^j(\mathbf{z}) = \sum_l \frac{\partial}{\partial f^l}\{f_n{}^j(\mathbf{f}(\mathbf{z}))\}\frac{\partial^2 f^l}{\partial z_i \partial z_s}$$
$$+\sum_l \sum_r \frac{\partial^2}{\partial f^l \partial f^r}\{f_n{}^j(\mathbf{f})\}\frac{\partial f^l}{\partial z_i}\frac{\partial f^r}{\partial z_s}.$$

Some further notation is necessary. Let us denote the covariance matrix of $\mathbf{Z}_1$ given that $\mathbf{Z}_0 = \mathbf{e}_j$ by $\mathbf{V}_j$, and the covariance matrix of

[2] In this chapter we use column vectors and prior matrix multiplication in order to be consistent with the Leslie notation of chapter 4. Row vectors are more usual in branching theory.

$\mathbf{Z}_n$ by $\mathbf{C}_n$. If all the $\{z_l\}$ are set equal to one in the above partial derivative, and the formula is re-arranged, we find that

$$\mathbf{C}_{n+1} = \mathbf{M}\mathbf{C}_n\mathbf{M}' + \sum_{l=1}^{k} \mathscr{E}(Z_{l,n})\,\mathbf{V}_l, \tag{8.4.10}$$

where $\mathbf{M}'$ denotes the transpose of the matrix $\mathbf{M}$. Equation (8.4.10) may be applied recursively to prove that

$$\mathbf{C}_n = \mathbf{M}^n\mathbf{C}_0\mathbf{M}'^n + \sum_{j=1}^{n} \mathbf{M}^{n-j}\left\{\sum_{l=1}^{k} \mathscr{E}(Z_{l,j-1})\,\mathbf{V}_l\right\}\mathbf{M}'^{n-j}. \tag{8.4.11}$$

For the basic multi-type process we are considering, the covariance matrix $\mathbf{C}_0$ will be identically zero. No further use will be made of equation (8.4.11) in this chapter, but equation (8.4.10) should be compared with Bartlett's equation (6.2.5).

As with the single-type process, we shall be concerned with the probability of extinction of the population, but it is necessary first to define a *singular* multi-type Galton–Watson process. This is a process in which every individual has exactly one child. We shall define q_j $(j = 1, 2, \ldots, k)$ to be the probability of extinction if initially there is one object of type j. The vector (q_j) will be denoted by $\mathbf{q}$ and a column vector of ones will be denoted by $\mathbf{1}$.

Theorem 8.4.1. Consider a non-singular multi-type Galton–Watson process with a positive regular[3] expectation matrix $\mathbf{M}$. The expectation matrix will have an algebraically simple, positive, dominant latent root λ_0. If $\lambda_0 \leqslant 1$, then $\mathbf{q} = \mathbf{1}$. If $\lambda_0 > 1$, then

$$0 \leqslant q_i < 1 \;(i = 1, 2, \ldots, k),$$

and $\mathbf{q}$ satisfies the equation

$$\mathbf{q} = \mathbf{f}(\mathbf{q}). \tag{8.4.12}$$

Proof. The theorem of Perron and Frobenius (theorem 4.3.1) guarantees that the expectation matrix will have a positive dominant latent root of algebraic multiplicity one. Furthermore, it is soon apparent that if the vector $\mathbf{q}$ lists the probabilities of extinction, it must be equal to $\mathbf{f}(\mathbf{q})$. We therefore examine the roots of equation (8.4.12).

Because the first- and second-order moments of $\mathbf{Z}_1$ are assumed to be finite, the first- and second-order left-hand derivatives of $\mathbf{f}(\mathbf{z})$ exist at $\mathbf{z} = \mathbf{1}$, and they are finite. Let us examine $\mathbf{f}(\mathbf{1}-\mathbf{h})$ where the elements of $\mathbf{h}$ are all positive and less than one. For simplicity, we shall

[3] *Positive regularity* is defined in section 4.3.

consider the two-type case, but the results and methods are quite general. According to Taylor's theorem,

$$\begin{aligned} f^1(1-h_1, 1-h_2) &= f^1(1,1) - h_1\frac{\partial f^1(1,1)}{\partial z_1} - h_2\frac{\partial f^1(1,1)}{\partial z_2} \\ &\quad + \tfrac{1}{2}h_1^2\frac{\partial^2}{\partial z_1^2} f^1(1-\theta h_1, 1-\theta h_2) \\ &\quad + h_1 h_2 \frac{\partial^2}{\partial z_1 \partial z_2} f^1(1-\theta h_1, 1-\theta h_2) \\ &\quad + \tfrac{1}{2}h_2^2\frac{\partial^2}{\partial z_2^2} f^1(1-\theta h_1, 1-\theta h_2), \end{aligned} \tag{8.4.13}$$

where $0 \leqslant \theta \leqslant 1$. A similar formula exists for $f^2(1-h_1, 1-h_2)$.

Let us consider first the case in which $\lambda_0 \leqslant 1$. The second-order derivatives are positive, and because $f^i(1,1)$ is equal to one, and $\partial f^j(1,1)/\partial z_i$ is equal to m_{ij}, we conclude that

$$\left.\begin{aligned} f^1(1-h_1, 1-h_2) &> 1 - h_1 m_{11} - h_2 m_{21}; \\ f^2(1-h_1, 1-h_2) &> 1 - h_1 m_{12} - h_2 m_{22}. \end{aligned}\right\} \tag{8.4.14}$$

These inequalities may be written in the vector form

$$\mathbf{1} - \mathbf{f}(\mathbf{1}-\mathbf{h}) < \mathbf{M}'\mathbf{h}, \tag{8.4.15}$$

where the inequality is interpreted as applying to each element.

Now the dominant latent root of $\mathbf{M}'$ is less than or equal to one, and it follows that at least one element of the vector $\mathbf{M}'\mathbf{h}$ must be less than or equal to the corresponding element of the vector $\mathbf{h}$. That is,

$$1 - f^j(\mathbf{1}-\mathbf{h}) < h_j \tag{8.4.16}$$

for some j. But we seek a vector $\mathbf{h}$ of positive elements less than one such that

$$\mathbf{1} - \mathbf{f}(\mathbf{1}-\mathbf{h}) = \mathbf{h},$$

and we conclude from inequality (8.4.16) that no such vector exists. At least one element h_j must be zero, and the probability of extinction when the initial ancestor is of that type is therefore one. Let us now remove the jth probability-generating function from consideration and examine the remaining $k-1$ functions. A series of inequalities like (8.4.15) is obtained, and the matrix $\mathbf{M}'$ is now without its jth row and jth column. The dominant latent root of this smaller matrix will still be less than or equal to one, and we finally conclude that all the elements of h must be zero. Thus when $\lambda_0 \leqslant 1$, the population will always become extinct whatever the type of initial ancestor.

The case $\lambda_0 > 1$ must now be considered, and we start with equation (8.4.13) and those similar to it. If the vector $\mathbf{h}\,(0 < h_j < 1)$ is selected so that it is proportional to the dominant latent vector of $\mathbf{M}'$, it is soon apparent that for sufficiently small $\mathbf{h}$, $f^j(\mathbf{1}-\mathbf{h})$ will be less than $1-h_j$. If there are non-zero probabilities of zero offspring, $f^j(\mathbf{1}-\mathbf{h})$ will be greater than $1-h_j$ where $\mathbf{h}$ is in the neighbourhood of $\mathbf{1}$. It is clear therefore that for equation (8.4.12) at least one root $\mathbf{q}$ exists which is subject to the inequality $\mathbf{0} \leqslant \mathbf{q} < \mathbf{1}$.

It remains to be proved that the obvious root $\mathbf{q} = \mathbf{1}$ is irrelevant, and that no element of $\mathbf{q}$ can be one. Let us imagine that $\mathbf{f}_n(\mathbf{0})$ is equal to $\mathbf{1}-\mathbf{h}$ where $\mathbf{h}$ is small. According to equation (8.4.4)

$$f_{n+1}{}^j(\mathbf{0}) = f^j(\mathbf{f}_n(\mathbf{0}))$$

and therefore, using equation (8.4.13), $f_{n+1}{}^j(\mathbf{0})$ will be *less* than $f_n{}^j(\mathbf{0})$ for at least one j. However it is known that $f_n{}^j(\mathbf{0})$ must be a monotonic increasing function of n, and we conclude that it must have a limit strictly below one. The root $\mathbf{q} = \mathbf{1}$ is therefore irrelevant.

To see that none of the $\{q_i\}$ can be equal to one, let us assume that the first t elements of $\mathbf{q}$ are one and that the remaining $k-t$ elements are less than one. Then

$$1 = f^i(1, \ldots, 1, q_{t+1}, \ldots, q_k) \quad (i = 1, \ldots, t).$$

These equations imply that the functions f^i $(i = 1, \ldots, t)$ are independent of $z_{t+1}, \ldots, z_k$. This is impossible due to the positive-regularity assumption, and it follows that all the elements of $\mathbf{q}$ must be less than one. The proof of the theorem is now complete.

The standard text on branching processes has been written by T. E. Harris (1963). The book contains a wealth of information, and the interested reader is referred to it. Two further theorems are relevant in the present chapter, and we therefore quote them. The first is a uniqueness theorem, and a proof is given on page 42 of T. E. Harris (1963).

Theorem 8.4.2. Consider a non-singular multi-type Galton–Watson process with a positive regular expectation matrix $\mathbf{M}$. If $\mathbf{q}_1$ is any vector in the unit cube other than $\mathbf{1}$, then

$$\lim_{n\to\infty} \mathbf{f}_n(\mathbf{q}_1) = \mathbf{q},$$

where $\mathbf{q}$ is the vector defined in theorem 8.4.1.

The other theorem will be of considerable interest in chapter 9, and a proof is given later in section 9.4.

Theorem 8.4.3. If $\mathbf{M}$ is positive regular and $\lambda_0 > 1$, then the vector random variable $\mathbf{Z}_n/\lambda_0^n$ converges with probability one to a random scalar multiple of the dominant latent vector of $\mathbf{M}$.

8.5 *Two examples*

For a certain one-type Galton–Watson process with mean m not equal to one, the probabilities $\{p_j\}$ $(j > 0)$ form a geometric series of the form

$$p_j = bc^{j-1} \quad (j = 1, 2, \ldots) \quad \text{and} \quad p_0 = 1 - \sum_{j=1}^{\infty} p_j. \qquad (8.5.1)$$

The constants b and c are positive and $b \leqslant 1 - c$. Determine the functional iterates $f_n(z)$, and the probability q of ultimate extinction.

The probability-generating function $f(z)$ takes the form

$$f(z) = \sum_{j=0}^{\infty} p_j z^j = 1 - \frac{b}{1-c} + \frac{bz}{1-cz}, \qquad (8.5.2)$$

and it is easy to see that the mean

$$m = f'(1) = \frac{b}{(1-c)^2}. \qquad (8.5.3)$$

The equation $f(z) = z$ has a positive root z_0 which is distinct from one, and this root is the probability of extinction when m is greater than one. When m is less than one, the probability of ultimate extinction is of course one, and the root z_0 lies outside the unit interval. It is easy to prove that

$$z_0 = \frac{1-b-c}{c(1-c)}. \qquad (8.5.4)$$

It remains to determine the form of the functional iterates. We first note that formula (8.5.2) may be written in the form

$$f(z) = 1 - m\left(\frac{1-z_0}{m-z_0}\right) + \frac{m\left(\dfrac{1-z_0}{m-z_0}\right)^2 z}{1 - \left(\dfrac{m-1}{m-z_0}\right) z},$$

and it may be proved by induction that

$$f_n(z) = 1 - m^n\left(\frac{1-z_0}{m^n-z_0}\right) + \frac{m^n\left(\dfrac{1-z_0}{m^n-z_0}\right)^2 z}{1 - \left(\dfrac{m^n-1}{m^n-z_0}\right) z}. \qquad (8.5.5)$$

This formula is valid whenever $m \neq 1$. The generating-functions used in this example are known as *fractional linear generating-functions*.

Branching process techniques find application in many different fields, and the second example, due to R. A. Fisher (1922), comes from genetics. The analysis is of historical interest in that Fisher appears to have rediscovered the one type branching process. We consider a large random-mating population in which all the individuals are of genotype AA, and assume that there is a small amount of mutation yielding the mutant a. If the average number of offspring per individual in the population is m, each gene will produce an average of m genes in the following generation. We shall assume that the number of genes produced by an individual gene is a Poisson random variable with mean m. A mutant a appears in the population. What is the probability that it does not become extinct?

The probability-generating function for the process is

$$f(z) = e^{-m} \sum_{j=0}^{\infty} z^j m^j / j! = e^{m(z-1)}. \tag{8.5.6}$$

If $m \leqslant 1$, we know from section 8.2 that the probability of ultimate extinction is one. If $m > 1$, the probability of extinction is less than one and we need to solve the equation $f(z) = z$, or

$$\log z = m(z-1). \tag{8.5.7}$$

Let us imagine that m is slightly greater than one and equal to $1+\delta$. In this case the appropriate root will be slightly less than one. If we write $1-y$ for z and expand $\log(1-y)$ in a power series, we find that

$$y + \tfrac{1}{2}y^2 + \tfrac{1}{3}y^3 \doteqdot (1+\delta)\,y. \tag{8.5.8}$$

Dividing throughout by y to remove the irrelevant root $y = 0$, and re-arranging, we obtain the quadratic equation

$$2y^2 + 3y - 6\delta \doteqdot 0,$$

with solution

$$y = -\tfrac{3}{4}\{1 - (1 + \tfrac{16}{3}\delta)^{\frac{1}{2}}\} \doteqdot 2\delta - \tfrac{8}{3}\delta^2. \tag{8.5.9}$$

If $m = 1.01$, for example, $\delta = 0.01$ and the probability of the continued survival of the mutant is 0.0197, confirming Fisher's calculation.

In this analysis, we have assumed a Poisson distribution. M. S. Bartlett (1955) gives a slightly more general result for the probability of ultimate extinction:

$$\Pr(\text{extinction}) \doteqdot \exp\{-2(m-1)/\sigma^2\}, \tag{8.5.10}$$

where m is the mean number of offspring and σ^2 is the variance in the number of offspring. The approximation is only valid when $m-1$ is rather small and positive.

An example of the multi-type branching process in a genetical context has been given by J. H. Pollard (1968*a*).

8.6 *Exercises*

1 Fit a geometric series to the probabilities (other than p_0) in Lotka's example (section 8.3), and use the results of section 8.5. to determine the probability of ultimate extinction.

2 For a certain one-type Galton–Watson process, the mean m is equal to one and the probabilities $\{p_j\}$ $(j > 0)$ form a geometric series of the form $p_j = bc^{j-1}$. Determine the functional iterates $f_n(z)$ and the probability q of ultimate extinction.

3 Use the methods of this chapter to analyse the population described in question 5 of chapter 6, and determine the probability of ultimate extinction if at time $t = 0$ there is exactly one individual and he is in age group 0.

4 What is the probability of ultimate extinction of the population of question 3 if the initial ancestor is in age group 1?

5 For a certain two-type Galton–Watson process,

$$f^1(z_1, z_2) = \frac{a+bz_1+cz_2}{d+ez_1+fz_2},$$

and

$$f^2(z_1, z_2) = \frac{g+hz_1+iz_2}{d+ez_1+fz_2}.$$

Find $f_2^1(z_1, z_2)$ and $f_2^2(z_1, z_2)$, and prove that the functional iterates f_n^1 and f_n^2 are all fractional linear.

6 Use the calculations of question 1 to compute the probability that the Lotka population will be extinct after 1, 2, ..., 10 generations. Equation (8.5.5) will be helpful.

7 Prove Bartlett's formula (8.5.10).

8 Use Bartlett's formula (8.5.10) to deduce the approximate probability of continued survival when the probability-generating function $f(z)$ is that of a Poisson distribution with mean $m = 1+\delta$. Compare this result with formula (8.5.9).

9
The stochastic version of Leslie's model

9.1 *Introduction*

The stochastic version of the continuous-time (Sharpe and Lotka) model, constructed by D. G. Kendall in 1948, was described in section 6.4. The theory led to an integral equation for the expectation behaviour which is identical with that of the deterministic theory, but Kendall's integral equations for the quadratic moments are difficult to solve, save in some unrealistic special cases. Kendall's continuous stochastic model was inspired by the earlier investigation by M. S. Bartlett (section 6.3) in which discrete time and discrete age groups were used. Bartlett's analysis was designed to yield asymptotic forms for the linear and quadratic moments when the age groups and time intervals were both made small, and so his work lies half-way between a discrete and a continuous formulation.

The exact stochastic analogue of Leslie's theory was published by J. H. Pollard, in 1966. In this model, fixed (not necessarily small) age-steps and time-steps are used, and although the theory can be regarded as a special case of the multi-type Galton–Watson process (section 8.4) the techniques developed and used by Pollard are different from those usually associated with branching processes, and well suited to population studies.

The fact that difficulties arise in the discrete formulation due to the effects of grouping was mentioned in section 4.1. When formulating discrete-time stochastic models, the main point to be borne in mind is that in general the effect on a Markov process of lumping states together is to cause the model to become non-Markovian. Usually, it is assumed that the underlying continuous-time model is Markovian, and *under this assumption* the discrete-time model cannot be Markovian. In this chapter, continuous-time processes are not considered, and the discrete-time processes are assumed to be Markovian. This means, of course, that other discrete-time models using different time units will be non-Markovian. However, for sufficiently small time units, the errors due to grouping should be small compared with those due to the other limitations of the model.

Certain relationships between the discrete-time models and the continuous-time models are derived in section 9.9.

9.2 *Some preliminary results*

In this chapter, we shall make extensive use of the *Kronecker product* (or *direct product*) of two matrices. Let $\mathbf{X} = (X_{ij})$ and $\mathbf{Y} = (Y_{ij})$ be matrices of dimension $l \times m$ and $r \times s$ respectively. Then the Kronecker product of $\mathbf{X}$ and $\mathbf{Y}$ is denoted by $\mathbf{X} \times \mathbf{Y}$ and is defined by

$$\mathbf{X} \times \mathbf{Y} = \begin{pmatrix} X_{11}\mathbf{Y} & X_{12}\mathbf{Y} & \dots & X_{1m}\mathbf{Y} \\ X_{21}\mathbf{Y} & X_{22}\mathbf{Y} & \dots & X_{2m}\mathbf{Y} \\ \vdots & \vdots & \vdots & \vdots \\ X_{l1}\mathbf{Y} & X_{l2}\mathbf{Y} & \dots & X_{lm}\mathbf{Y} \end{pmatrix}, \tag{9.2.1}$$

which is a matrix of dimension $lr \times ms$.

Lemma 9.2.1. If the Leslie matrix $\mathbf{A}$ has latent roots $\{\lambda_i\}$ $(i = 0, 1, \dots, k)$, the matrix $\mathbf{A} \times \mathbf{A}$ has latent roots $\{\lambda_i \lambda_j\}$ $(i, j = 0, 1, \dots, k)$. If the matrix $\mathbf{A}$ is positive regular,[1] and its algebraically simple, positive, strictly dominant, latent root is λ_0 with corresponding right and left latent vectors $\mathbf{x}$ and $\mathbf{y}'$, then the matrix $\mathbf{A} \times \mathbf{A}$ is positive regular and its algebraically simple, positive, strictly dominant latent root is λ_0^2, with corresponding right and left latent vectors $\mathbf{x} \times \mathbf{x}$ and $\mathbf{y}' \times \mathbf{y}'$ (i.e. the Kronecker products as defined above).

Proof. The following properties of the Kronecker product follow readily from the definition.

(i) If $\mathbf{X}$, $\mathbf{Y}$, $\mathbf{U}$ and $\mathbf{V}$ are matrices of any size and shape, and providing that the ordinary matrix multiplications are possible, $(\mathbf{X} \times \mathbf{Y})(\mathbf{U} \times \mathbf{V}) \equiv (\mathbf{XU}) \times (\mathbf{YV})$.

(ii) If $\mathbf{X}$ and $\mathbf{Y}$ are square and non-singular, then so is $\mathbf{X} \times \mathbf{Y}$, and $(\mathbf{X} \times \mathbf{Y})^{-1} \equiv \mathbf{X}^{-1} \times \mathbf{Y}^{-1}$.

From the theory of matrices,[2] it is known that there is a non-singular matrix $\mathbf{H}$ such that $\mathbf{HAH}^{-1} = \mathbf{Z}$, where $\mathbf{Z}$ consists of diagonally arranged blocks like

$$\mathbf{Z}_{(r)} = \begin{pmatrix} \lambda_{(r)} & 1 & & & \\ & \lambda_{(r)} & 1 & & \\ & & \ddots & \ddots & \\ & & & \ddots & 1 \\ & & & & \lambda_{(r)} \end{pmatrix}.$$

[1] *Positive regularity* is defined in section 4.3.

[2] See, for example, H. W. Turnbull and A. C. Aitken, 1951, 58–70.

Now

$$(\mathbf{H}\times\mathbf{H})(\mathbf{A}\times\mathbf{A})(\mathbf{H}\times\mathbf{H})^{-1} = (\mathbf{H}\times\mathbf{H})(\mathbf{A}\times\mathbf{A})(\mathbf{H}^{-1}\times\mathbf{H}^{-1})$$
$$= (\mathbf{HAH}^{-1})\times(\mathbf{HAH}^{-1}),$$

and this matrix has elements $\{\lambda_i\lambda_j\}$ on the diagonal, and no non-zero elements below the diagonal. The latent roots of $\mathbf{A}\times\mathbf{A}$ are the same as those of $(\mathbf{H}\times\mathbf{H})(\mathbf{A}\times\mathbf{A})(\mathbf{H}\times\mathbf{H})^{-1}$, and we conclude that the latent roots of

$$\mathbf{A}\times\mathbf{A} \quad \text{are} \quad \{\lambda_i\lambda_j\} \quad (i,j = 0, 1, \ldots, k).$$

It follows immediately from the definition of a Kronecker product that $\mathbf{A}\times\mathbf{A}$ is positive regular if $\mathbf{A}$ is so. If $\mathbf{A}$ is positive regular, it will have a strictly dominant, algebraically simple, positive latent root λ_0, by the theorem of Perron and Frobenius. Alternatively, it is known from section 4.4 that the Leslie matrix has a strictly dominant, algebraically simple, positive latent root provided two consecutive F_j are non-zero and we assume this condition to be true. It is clear therefore that $\mathbf{A}\times\mathbf{A}$ has a strictly dominant, algebraically simple, positive latent root λ_0^2.

The column eigenvector of $\mathbf{A}$ corresponding to the dominant latent root λ_0 is $\mathbf{x}$, and we note that

$$(\mathbf{A}\times\mathbf{A})(\mathbf{x}\times\mathbf{x}) = (\mathbf{Ax})\times(\mathbf{Ax}) = (\lambda_0\mathbf{x})\times(\lambda_0\mathbf{x}) = \lambda_0^2(\mathbf{x}\times\mathbf{x}).$$

We conclude that $\mathbf{x}\times\mathbf{x}$ is the column latent vector of $\mathbf{A}\times\mathbf{A}$ corresponding to the dominant latent root λ_0^2. Similarly, $\mathbf{y}'\times\mathbf{y}'$ is the row latent vector of $\mathbf{A}\times\mathbf{A}$ corresponding to the dominant latent root λ_0^2, and it should be noted that $(\mathbf{y}'\times\mathbf{y}')(\mathbf{x}\times\mathbf{x}) = 1$ because $\mathbf{y}'\mathbf{x} = 1$ (section 4.4).

9.3 *The stochastic population model*

The female population only is considered, at discrete points of time $t = 0, 1, 2, \ldots$, and the $k+1$ age groups $0-$, $1-$, ..., $k-$ mentioned in connection with Leslie's model are again used. Changes in the male population structure are assumed to be consistent with the assumption of constant fertility measures. For this stochastic model, F_x is defined to be the probability that a female in age group $x-$ at time t will give birth to a *single* daughter during the time interval $(t, t+1)$ and that this daughter will be alive at time $t+1$ to be enumerated in age group $0-$. Multiple births will be discussed in section 9.5. P_x is defined as the probability of a female in age group x

at time t surviving to be in age group $x+1$ at time $t+1$. Quantities $Q_x = 1-P_x$ and $G_x = 1-F_x$ are also defined.

The number of females in the age group $x-$ at time t is a random variable $n_{x,t}$ with expected value $e_{x,t}$ and variance $C^{(t)}_{x,x}$. The covariance $\operatorname{Cov}(n_{x,t}, n_{y,t})$ is denoted by $C^{(t)}_{x,y}$. The number of females in the age group $0-$ at time t whose mothers were aged x at the time of the birth is a random variable $n^{(x)}_{0,t}$. (That is, their mothers were among the females enumerated by $n_{x,t-1}$.) Obviously,

$$n_{0,t} = \sum_x n^{(x)}_{0,t}. \tag{9.3.1}$$

The stochastic model is set up as follows. Consider the $n_{x,t}$ females aged x at time t. Each of them has a fixed probability P_x of surviving the unit time interval, and they are assumed independent. Hence $n_{x+1,t+1}$ is a conditional binomial random variable $B(n_{x,t}, P_x)$ conditional on $n_{x,t}$. Similarly, each of the $n_{x,t}$ females has a fixed probability F_x of contributing a single[3] daughter in the age group $0-$ at time $t+1$, and births and deaths are assumed independent.[4]

Using equation (6.5.1) of lemma 6.5.1, we see that the Leslie recurrence equations (4.2.1) are still true when the deterministic values $\{n_{x,t}\}$ are replaced by the stochastic expectations $\{e_{x,t}\}$. Thus

$$\mathbf{e}_{t+1} = \mathbf{A}\mathbf{e}_t. \tag{9.3.2}$$

The following equations are obtained using equations (6.5.2), (6.5.3) and (6.5.4) of the same lemma:

$$C^{(t+1)}_{x+1,x+1} = P_x^2 C^{(t)}_{x,x} + P_x Q_x e_{x,t} \quad (x \geqslant 0), \tag{9.3.3}$$

$$C^{(t+1)}_{x+1,y+1} = P_x P_y C^{(t)}_{x,y} \quad (x, y \geqslant 0, x \neq y), \tag{9.3.4}$$

$$\operatorname{Cov}(n^{(x)}_{0,t+1}, n_{x+1,t+1}) = F_x P_x C^{(t)}_{x,x} \quad (x \geqslant 0), \tag{9.3.5}$$

$$\operatorname{Cov}(n^{(x)}_{0,t+1}, n_{y+1,t+1}) = F_x P_y C^{(t)}_{x,y} \quad (x \neq y), \tag{9.3.6}$$

$$\operatorname{Cov}(n^{(x)}_{0,t+1}, n^{(y)}_{0,t+1}) = F_x F_y C^{(t)}_{x,y} \quad (x \neq y), \tag{9.3.7}$$

$$\operatorname{Var}(n^{(x)}_{0,t+1}) = F_x^2 C^{(t)}_{x,x} + F_x G_x e_{x,t} \quad (x \geqslant 0). \tag{9.3.8}$$

Now by definition

$$n_{0,t+1} = \sum_{x=0}^{k} n^{(x)}_{0,t+1},$$

[3] Multiple births are accounted for in section 9.5.

[4] It is not necessary to assume that births and deaths are independent. This becomes apparent after section 9.7. The assumption of independence simplifies the present analysis.

and we deduce that

$$C_{0,0}^{(t+1)} = \sum_{x=0}^{k} (F_x{}^2 C_{x,x}^{(t)} + F_x G_x e_{x,t}) + \sum\sum_{x \neq y} F_x F_y C_{x,y}^{(t)}; \quad (9.3.9)$$

$$C_{0,y+1}^{(t+1)} = \sum_{\text{all } x} F_x P_y C_{x,y}^{(t)}. \quad (9.3.10)$$

Equations (9.3.2), (9.3.3), (9.3.4), (9.3.9) and (9.3.10) completely define the recurrence relations for the means, variances and covariances. Furthermore, they are linear recurrence relations, and they may be written in the form

$$\begin{pmatrix} \mathbf{e}_{t+1} \\ \mathbf{C}(t+1) \end{pmatrix} = \begin{pmatrix} \mathbf{A} & \mathbf{0} \\ \mathbf{D} & \mathbf{A} \times \mathbf{A} \end{pmatrix} \begin{pmatrix} \mathbf{e}_t \\ \mathbf{C}(t) \end{pmatrix}, \quad (9.3.11)$$

where $\mathbf{A}$ is the $(k+1) \times (k+1)$ Leslie matrix. The vector $\mathbf{e}_t$ is the vector of expectations at time t. The vector $\mathbf{C}(t)$ has as its elements the variances and covariances $\{C_{i,j}^{(t)}\}$ and these are listed in dictionary order according to the subscripts i and j. For $i \neq j$, $C_{i,j}^{(t)}$ and $C_{j,i}^{(t)}$ are both listed. It is of interest to note that by the repeated application of equation (9.3.11),

$$\mathbf{C}(n) = (\mathbf{A} \times \mathbf{A})^n \, \mathbf{C}(0) + \sum_{j=1}^{n} (\mathbf{A} \times \mathbf{A})^{n-j} \, (\mathbf{D}\mathbf{e}_{j-1}). \quad (9.3.12)$$

This is another form of equation (8.4.11).

If we now write $\mathbf{C}[n]$ for the vector of non-central quadratic moments at time n,

$$\mathbf{C}[n] = \mathbf{C}(n) + (\mathbf{e}_n \times \mathbf{e}_n) = \mathbf{C}(n) + (\mathbf{A} \times \mathbf{A})^n \, (\mathbf{e}_0 \times \mathbf{e}_0).$$

Hence

$$\mathbf{C}[n] = (\mathbf{A} \times \mathbf{A})^n \, \mathbf{C}[0] + \sum_{j=1}^{n} (\mathbf{A} \times \mathbf{A})^{n-j} \, (\mathbf{D}\mathbf{e}_{j-1}). \quad (9.3.13)$$

This is equation (4.3) on page 37 of T. E. Harris (1963), but written in a rather different form. It was derived here for a special case of the multi-type Galton–Watson process. Note that the central and non-central quadratic moments obey the same recurrence equation. Only the initial conditions are different.

9.4 *The asymptotic behaviour of the means and variances*

The asymptotic behaviour of the means is already known from the analysis of Leslie's model. Equation (4.5.3) give us these results. It is easy to write down the eigenvectors $\mathbf{x}$ and $\mathbf{y}'$, as was shown in section 4.4. To determine the asymptotic behaviour of the central quadratic moments, the dominant latent root of the matrix in equation (9.3.11) is required. The characteristic equation may be written in the form

$$|\mathbf{A}-\lambda\mathbf{I}|\;|\mathbf{A}\times\mathbf{A}-\lambda\mathbf{I}\times\mathbf{I}| = 0, \tag{9.4.1}$$

and we see that the matrix has as its latent roots all the latent roots of $\mathbf{A}$ together with all the latent roots of $\mathbf{A}\times\mathbf{A}$. By lemma 9.2.1, the dominant latent root of $\mathbf{A}\times\mathbf{A}$ is λ_0^2 and it has algebraic multiplicity one. There are therefore three cases to consider:

(i) $\lambda_0 > 1$. In this case $\mathbf{A}\times\mathbf{A}$ contributes the algebraically simple dominant latent root λ_0^2, with latent column vector

$$\begin{pmatrix}\mathbf{0}\\ \mathbf{x}\times\mathbf{x}\end{pmatrix}.$$

(ii) $\lambda_0 = 1$. In this case there is a pair of dominant latent roots and each is equal to one.

(iii) $\lambda_0 < 1$. In this case $\mathbf{A}$ contributes the algebraically simple dominant latent root λ_0, with latent column vector

$$\begin{pmatrix}\mathbf{x}\\ \mathbf{x}^*\end{pmatrix}, \quad \text{say.}$$

Case (i) will be discussed in considerable detail; case (ii) is of no practical significance, and case (iii) is difficult to analyse. In case (iii), we do have however

$$\mathbf{x}^* = (\lambda_0\mathbf{I}\times\mathbf{I}-\mathbf{A}\times\mathbf{A})^{-1}\,\mathbf{D}\mathbf{x}, \tag{9.4.2}$$

and the corresponding normalized left latent vector of the moment matrix is easily shown to be $(\mathbf{y}' \quad \mathbf{0})$. Applying lemma 4.5.1,

$$\lambda_0^{-n}\begin{pmatrix}\mathbf{A} & \mathbf{0}\\ \mathbf{D} & \mathbf{A}\times\mathbf{A}\end{pmatrix}^n = \begin{pmatrix}\mathbf{x}\mathbf{y}' & \mathbf{0}\\ \mathbf{x}^*\mathbf{y}' & \mathbf{0}\end{pmatrix}+o(1), \tag{9.4.3}$$

and the following asymptotic results are obtained:

$$\left.\begin{aligned}\mathbf{e}_t/\lambda_0^t &\simeq (\mathbf{y}'\mathbf{e}_0)\,\mathbf{x};\\ \mathbf{C}(t)/\lambda_0^t &\simeq (\mathbf{y}'\mathbf{e}_0)\,(\lambda_0\mathbf{I}\times\mathbf{I}-\mathbf{A}\times\mathbf{A})^{-1}\,\mathbf{D}\mathbf{x}.\end{aligned}\right\} \tag{9.4.4}$$

Let us now continue with case (i). The dominant[5] latent root λ_0^2 is of algebraic multiplicity one, and the corresponding right latent

[5] Note that the recurrence matrix in equation (9.3.11) is *not* positive regular, and that there are at least two positive latent roots (λ_0 and λ_0^2).

vector is given above. Let the normalized dominant left latent vector be $(\mathbf{u}' \quad \mathbf{v}')$. Then it is easy to prove that

$$\left.\begin{aligned} \mathbf{u}' &= (\mathbf{y}' \times \mathbf{y}')\mathbf{D}(\lambda_0^2\mathbf{I}-\mathbf{A})^{-1}; \\ \mathbf{v}' &= \mathbf{y}' \times \mathbf{y}'. \end{aligned}\right\} \tag{9.4.5}$$

Applying lemma 4.5.1,

$$\lambda_0^{-2n}\begin{pmatrix} \mathbf{A} & \mathbf{0} \\ \mathbf{D} & \mathbf{A}\times\mathbf{A}\end{pmatrix}^n = \begin{pmatrix} \mathbf{0} & \mathbf{0} \\ (\mathbf{x}\times\mathbf{x})\,\mathbf{u}' & (\mathbf{x}\mathbf{y}')\times(\mathbf{x}\mathbf{y}')\end{pmatrix} + o(1), \tag{9.4.6}$$

and the following asymptotic results are obtained:

$$\left.\begin{aligned} \mathbf{e}^t/\lambda_0^{2t} &\simeq \mathbf{0} \text{ (as it should!)}; \\ \mathbf{C}(t)/\lambda_0^{2t} &\simeq \{\mathbf{u}'\mathbf{e}_0 + (\mathbf{y}'\times\mathbf{y}')\,\mathbf{C}(0)\}\,\mathbf{x}\times\mathbf{x}. \end{aligned}\right\} \tag{9.4.7}$$

It is interesting to note that these results imply immediately that for $\lambda_0 > 1$, and large t, the correlation between $n_{i,t}$ and $n_{j,t}$ is asymptotically one. This fact suggests that the vector random variable $\mathbf{n}_t$ converges in some sense to a random multiple of a fixed vector – the stable age distribution vector. The following theorem, due to C. J. Everett and S. Ulam (1948) and T. E. Harris (1951), describes the behaviour of $\mathbf{n}_t$ precisely.

Theorem 9.4.1. If $\mathbf{A}$ is positive regular and $\lambda_0 > 1$, then the vector random variable $\mathbf{n}_t/\lambda_0^t$ converges with probability one to a random scalar multiple of the unique stable age distribution $\mathbf{x}$.

Proof. To prove this, it is only necessary to apply theorem 9.2 on page 44 of Harris (1963), and to note that his $\mathbf{M}'$ is in fact equal to the Leslie matrix $\mathbf{A}$. It is possible to give a direct proof, however.

From lemma 4.5.1 we know that

$$\begin{aligned} \mathbf{e}_t &= \alpha\lambda_0^t\mathbf{x} + O(t^{k-1}|\lambda_1|^t); \\ \mathbf{C}(t) &= \beta\lambda_0^{2t}\mathbf{x}\times\mathbf{x} + O(t^{k^2+3k}|\lambda_1\lambda_0|^t). \end{aligned}$$

A set of random variables $\{X_{i,t}\}$ is defined as follows:

$$X_{i,t} = (n_{i,t} - \alpha\lambda_0^t x_i)/(\beta^{\frac{1}{2}}\lambda_0^t x_i).$$

Then for large t, it may be shown that

$$\mathscr{E}(X_{i,t} - X_{j,t})^2 = O(t^{k^2+3k}|\lambda_1/\lambda_0|^t) + O(t^{2k-2}|\lambda_1/\lambda_0|^{2t}).$$

Writing Z_n for $X_{i,t+n} - X_{j,t+n}$, we have

$$\lim_{n\to\infty} \mathscr{E}Z_n^2 = 0 \quad\text{and}\quad \sum_{n=1}^{\infty} \mathscr{E}Z_n^2 < \infty.$$

Z_n therefore converges to zero with probability one (E. Parzen, 1960, p. 416). Hence $\mathbf{n}_t/\lambda_0{}^t$ converges with probability one to the stable age distribution. This completes the proof of theorem 9.4.1. The proof of theorem 8.4.3 for the general multi-type process also follows these lines.

9.5 *The effect of multiple births*

So far in this analysis, it has been assumed that a female aged x can produce at most one daughter in the unit time interval. The effect of multiple births on the theory will now be investigated.

Let $F_{x,j}$ denote the probability of a female in age group $x-$ at time t contributing exactly j daughters to age group $0-$ at time $t+1$. It is convenient to define

$$F_x = \sum_j jF_{x,j} \quad \text{and} \quad G_x = 1-F_x.$$

Lemma 9.5.1. Let M be a random variable taking non-negative integral values and having mean μ and variance σ^2. Let

$$\{M_j'\} \quad (j = i, 2, \ldots, n)$$

be random variables having the conditional multinomial distribution Mult $(M; p_1, p_2, \ldots, p_n)$ conditional on M. Then the covariance of M_1' and M_2' is given by

$$\text{Cov}\,(M_1', M_2') = p_1p_2(\sigma^2-\mu). \tag{9.5.1}$$

The elementary proof is similar to that given for lemma 6.5.1.

Let $n_{0,t+1}^{(x,j)}$ be the random variable representing the number of females in age group $0-$ at time $(t+1)$, whose mothers were aged x at the time of their birth, and who were the result of a confinement yielding exactly j daughters at time $t+1$.

Obviously

$$n_{0,t+1} = \sum_j \sum_{x=0}^{k} n_{0,t+1}^{(x,j)}. \tag{9.5.2}$$

The modified stochastic model is fundamentally the same as that described in section 9.3. In fact the only difference is that the $\{(n_{0,t+1}^{(x,j)'})/j\}$ $(j = 1, 2, 3, \ldots)$ are multinomial random variables conditional on $n_{x,t}$, whereas previously $n_{0,t+t}^{(x)}$ was a binomial variable conditional on $n_{x,t}$.

Let us now investigate the effect of this modification of the stochastic model on the basic recurrence relation (9.3.11). It is immediately obvious that the expectations are still given by the Leslie equations (9.3.2), although F_x has now an expectation definition rather than a probability definition. It is also obvious that the recurrence relation for $C^{(t)}_{x,y}$ is unchanged except perhaps when either or both of x and y are zero. However, the recurrence relation for $C^{(t+1)}_{0,y+1}$ is unchanged by an expectation definition of F_x, because

$$\begin{aligned} C^{(t+1)}_{0,y+1} &= \operatorname{Cov}\left(\sum_j \sum_{x=0}^{k} n^{(x,j)}_{0,t+1}, n_{y+1,t+1}\right) \\ &= \sum_j \sum_{x=0}^{k} \operatorname{Cov}(n^{(x,j)}_{0,t+1}, n_{y+1,t+1}) \\ &= \sum_{x=0}^{k} \sum_j jF_{x,j} P_y C^{(t)}_{x,y} \\ &= \sum_{x=0}^{k} F_x P_y C^{(t)}_{x,y}. \qquad \text{(lemma 6.5.1)} \end{aligned}$$

There is now only $C^{(t+1)}_{0,0}$ to consider. After some straightforward but slightly tedious algebra, we find that

$$\left.\begin{aligned} C^{(t+1)}_{0,0} &= \sum_{x=0}^{k} \sum_{y=0}^{k} F_x F_y C^{(t)}_{x,y} + \sum_{x=0}^{k} \phi_x e_{x,t} \\ \text{where} \qquad \phi_x &= \sum_j (j-F_x)^2 F_{x,j}. \end{aligned}\right\} \qquad (9.5.3)$$

The elements $\{F_x G_x\}$ in the recurrence submatrix $\mathbf{D}$ are replaced by the elements $\{\phi_x\}$. This is the only alteration necessary. It is of interest to note that the eigenvalues of the recurrence matrix are unchanged, and the basic asymptotic results of section 9.4 are still true.

9.6 *An experiment using the stochastic model*

A Monte Carlo experiment using the model of section 9.3 was performed on TITAN, the computer of the Cambridge Computer Laboratory. This was done for the following reasons:

(i) to observe one particular outcome of the stochastic process and see how well the behaviour of the stochastic model is summarized by what is known about its quadratic moments;

(ii) to check the theoretical asymptotic results of section 9.4 in an actual numerical example; and

(iii) to look for any other interesting features in the results of the experiment.

The experiment was performed as follows: from data available about the Australian female population in 1960, approximate values of $\{F_i\}$ and $\{P_i\}$ were calculated for all ages from 0 to 59. The unit interval of time selected was one year. The experiment was *not* intended to be a prediction of the future Australian female population, and indeed it did not take into account many important factors (e.g. immigration). The data were used merely because they were convenient. The effect of multiple births was ignored.

In the experiment pseudo-random normal deviates[6] were generated by a suitable subroutine, and a normal approximation to the binomial distribution was used. This was reasonable because of the large numbers involved. The theoretical mean/variance/covariance vector was updated after each yearly step, in parallel with the Monte Carlo experiment. It is important to note that the Monte Carlo experiment and the theoretical mean/variance/covariance calculations were carried out *completely* independently, except of course in the initial year 1960. Some of the results obtained were output at intervals in the format of table 9.6.1. The theoretical matrix representing the process was enormous, but because most of the elements were zero, the computations were readily handled by TITAN when suitably programmed.

In table 9.6.1, the column headed 'actual' contains the numbers obtained from the Monte Carlo experiment. The figures in the columns headed 'expected' and 'variance' are theoretical values for the means and variances obtained using the recurrence relation (9.3.11). The figures in the column headed 'norm. devn.' represent the standardized deviation of the Monte Carlo value from the theoretical mean.

The following were the results of the experiment:

(*a*) It was found that over a long period of time (250 years) the standardized deviations of the Monte Carlo experimental values from the theoretical means were mainly in the range $(-3, 3)$ with occasional values outside this range. This is to be expected, although it should be noted that, because of the high positive correlation, the number of degrees of freedom between all the variables involved in the experiment is much less than 250×60.

(*b*) It is known from section 9.4 that when $\lambda_0 > 1$,

$$\lim_{t\to\infty} e_{j,t+1}/e_{j,t} = \lambda_0 \quad \text{and} \quad \lim_{t\to\infty} C_{i,j}^{(t+1)}/C_{i,j}^{(t)} = \lambda_0^2.$$

Approximate values of λ_0 and λ_0^2 can be obtained therefore by calculating these ratios for large values of t. For $t = 250$, the expectations

[6] See exercise 1 of section 9.11.

TABLE 9.6.1. *Results of the Monte Carlo experiment*

Age	Actual	Expected	Variance	Norm. devn.
		Year 2210		
0	9,341,453	9,338,791	63,254,673	0.33
5	8,338,516	8,331,608	51,143,933	0.97
10	7,609,272	7,607,037	43,203,020	0.34
15	6,955,804	6,949,451	36,575,368	1.05
20	6,343,290	6,343,082	30,954,317	0.04
25	5,787,151	5,786,012	26,200,367	0.22
30	5,276,824	5,274,404	22,176,170	0.51
35	4,802,855	4,800,826	18,744,253	0.47
40	4,355,119	4,357,383	15,791,753	−0.57
45	3,938,210	3,936,306	13,225,115	0.52
50	3,530,432	3,529,092	10,960,105	0.40
55	3,133,078	3,130,023	8,942,916	1.02
		Year 2211		
0	9,508,632	9,505,590	65,388,834	0.38
5	8,489,026	8,480,387	52,857,412	1.19
10	7,746,167	7,742,867	44,641,333	0.49
15	7,072,937	7,073,566	37,784,272	−0.10
20	6,457,004	6,456,402	31,969,728	0.11
25	5,890,323	5,889,380	27,053,629	0.18
30	5,370,942	5,368,591	22,893,122	0.49
35	4,887,074	4,886,521	19,344,997	0.13
40	4,435,979	4,435,180	16,292,343	0.20
45	4,007,306	4,006,639	13,639,127	0.18
50	3,595,328	3,592,176	11,298,875	0.94
55	3,187,583	3,185,936	9,215,682	0.54

Note. The values for every individual year of age were output by the computer. This table gives every fifth entry.

ratios are almost independent of j, and yield $\lambda_0 \doteqdot 1.01786$. However, with this value of t, the variances ratios still depend upon j, and a much larger value of t is necessary to be able to calculate λ_0^2 with reasonable accuracy from the variances ratios. In fact, the variances ratios for $t = 250$ are approximately 1.033 compared with

$$\lambda_0^2 = (1.01786)^2 = 1.03605.$$

These calculations are given in table 9.6.2.

(*c*) It is known from theorem 9.4.1, that when $\lambda_0 > 1$, the normalized deviations should be independent of t and j for large t. It is clear from table 9.6.1 that for $t = 250$ this is not so. However, it was observed under (*b*) that the variances ratios are not close to λ_0^2 for $t = 250$, and hence the group random variables at time $t = 250$ have

TABLE 9.6.2. *Variances and means ratios*

Age	Variances ratio 2211/2210	Means ratio 2211/2210
0	1.03373	1.017860
1	1.03365	1.017860
2	1.03361	1.017859
3	1.03358	1.017858
4	1.03354	1.017857
5	1.03350	1.017857
6	1.03346	1.017856
7	1.03342	1.017856
8	1.03338	1.017855
9	1.03333	1.017855
10	1.03329	1.017855
11	1.03324	1.017856
12	1.03319	1.017856
13	1.03314	1.017857
14	1.03310	1.017858
15	1.03305	1.017859
16	1.03300	1.017860
17	1.03295	1.017861
18	1.03290	1.017863
19	1.03285	1.017864

a distribution considerably different from their asymptotic distribution. At any rate most of the normalised deviations have the same sign! A record of the signs of the deviations at intervals of 20 years is given in table 9.6.3. If the experiment were continued, all the normalized deviations would ultimately have the same sign, and all would approach the same value for all j.

(*d*) Theorem 9.4.1 also tells us that for large t, an estimate of λ_0 will be given by the ratio of the total actual population at time $t+1$ to the total actual population at time t. This ratio was calculated for the years 2210 and 2211 and an estimate 1.01785 was obtained, which is close to the value 1.01786 based on the rate of increase of $e_{j,t}$.

(*e*) The smallness of the variances in table 9.6.1 should be noted. Consider age group 30 in year 2210; no demographer would predict an expected number of 5,274,404 with a standard deviation of 4,709. This point will be mentioned again in section 9.8.

TABLE 9.6.3. *Signs of deviations*

Year	Number of negative deviations
1960	(30)
1980	26
2000	24
2020	21
2040	20
2060	17
2080	15
2100	19
2120	15
2140	13
2210	4

Note. Initially all deviations are zero. The figure 30 in parentheses is half of 60, the number of age groups.

9.7 *Generalization to higher-order moments*

It is possible to generalize the results of the earlier sections of this chapter to deal with the higher-order moments about the origin in any multi-type Galton–Watson process. Although the general results are of little practical value, the techniques are very useful computationally for general multi-type processes when quadratic moments are required. Let us look at such a process more closely.

For each individual of type j at time t, there are several possible alternatives during the time interval $(t, t+1)$. For example, in the population model of section 9.3, there are four possibilities for each individual in each age group:

(i) Female aged x at time t survives to be aged $x+1$ at time $t+1$ having given birth to no daughters who survive to be age 0 at time $t+1$. The probability is $P_x(1-F_x)$, because births and deaths are assumed independent.

(ii) Female aged x at time t survives to be aged $x+1$ at time $t+1$ having given birth to a single daughter who is aged 0 at time $t+1$. The probability is $P_x F_x$.

(iii) Female aged x at time t dies during the unit time interval without leaving a surviving daughter aged 0 at time $t+1$. The probability is $(1-P_x)(1-F_x)$.

(iv) Female aged x at time t dies during the unit time interval leaving a single surviving daughter aged 0 at time $t+1$. The probability is $(1-P_x) F_x$.

Corresponding to each of these possible alternatives for each individual, there is a score to type i at time $t+1$. In the above population model, for example, the scores allotted to the various types at time $t+1$ are as follows for the four alternatives for each age group.

(i) Age group $x+1$ is allotted a score of unity; all the other age groups are allotted scores of zero.

(ii) Age groups $x+1$ and 0 are each allotted scores of unity; all other age groups are allotted scores of zero.

(iii) All age groups are allotted scores of zero.

(iv) Age group 0 is allotted a score of unity; all the other age groups are allotted scores of zero.

This method of breaking up the process leads us to consider (in addition to the random variables at times t and $t+1$) the random variables representing the numbers of females following the different alternatives. These random variables will be called *intermediary random variables.*

To simplify the discussion, a two-type Galton–Watson process will be considered. The number of individuals of type j at time t is represented by a random variable $n_j(t)$ $(j = 1, 2; t = 0, 1, 2, \ldots)$. Let the individual of type 1 have three possible alternatives in the unit time interval with probabilities 0.2, 0.3, and 0.5; and those of type 2, two possible alternatives with probabilities 0.25 and 0.75. The intermediary random variables will be denoted by $\{y_i(t)\}$ $(i = 1, 2, \ldots, 5)$. Then

$$y_1(t) + y_2(t) + y_3(t) = n_1(t),$$

and

$$y_4(t) + y_5(t) = n_2(t).$$

Thus the random variables $y_1(t)$, $y_2(t)$ and $y_3(t)$ are conditional multinomial variables conditional on $n_1(t)$. Similarly $y_4(t)$ and $y_5(t)$ are conditional multinomial (binomial) random variables conditional on $n_2(t)$. It is clear that the random variables $\{n_j(t+1)\}$ are merely linear transformations of the random variables $\{y_i(t)\}$. The diagram in figure 9.7.1 represents this process. Certain linear transformation constants have been chosen to complete the specification of the process, and these are given in figure 9.7.1.

Four observations are now made:

1. The conditional multinomial probabilities and the linear transformation constants completely specify the multi-type Galton–Watson process.[7]

2. Consider a linear transformation of a vector random variable

[7] The usual approach to a multi-type Galton–Watson process is given in section 8.4.

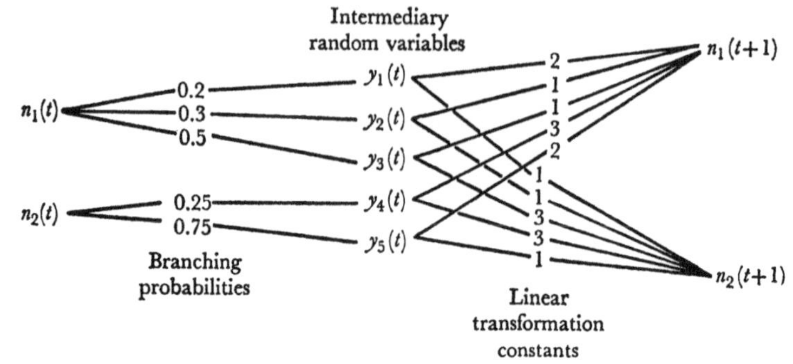

Figure 9.7.1. Diagrammatic representation of a simple two-type Galton–Watson process.

Y to another vector random variable **X** given by $\mathbf{X} = \mathbf{QY}$, where the matrix **Q** need not necessarily be square. All the third-order moments of **X** *about the origin* are listed in dictionary order in the vector

$$\mathscr{E}\mathbf{X}\times\mathbf{X}\times\mathbf{X} = \mathscr{E}(\mathbf{QY})\times(\mathbf{QY})\times(\mathbf{QY}) = (\mathbf{Q}\times\mathbf{Q}\times\mathbf{Q})\,\mathscr{E}\mathbf{Y}\times\mathbf{Y}\times\mathbf{Y}.$$

It follows that all the first-, second- and third-order moments about the origin (listed in dictionary order and in increasing degree) are connected by the matrix

$$\mathbf{T} = \begin{pmatrix} \mathbf{Q} & & \\ & \mathbf{Q}\times\mathbf{Q} & \\ & & \mathbf{Q}\times\mathbf{Q}\times\mathbf{Q} \end{pmatrix}.$$

The generalization to fourth- and higher-order moments is obvious. In the present context, **Q** will have non-negative integral elements. For the two-type example depicted in figure 9.7.1,

$$\mathbf{Q} = \begin{pmatrix} 2 & 1 & 1 & 3 & 2 \\ 1 & 1 & 3 & 3 & 1 \end{pmatrix}.$$

3. If the column vector of conditional multinomial probabilities for type j is $\mathbf{p}^{(j)}$, and we define

$$\mathbf{P} = \begin{pmatrix} \mathbf{p}^{(1)} & & & \\ & \mathbf{p}^{(2)} & & \\ & & \mathbf{p}^{(3)} & \\ & & & \ddots \end{pmatrix},$$

then the first-, second- and third-order falling factorial moments of the intermediary random variables and those of the random variables at time t (listed in dictionary order and in increasing degree) are connected by the matrix

$$\mathbf{B} = \begin{pmatrix} \mathbf{P} & & \\ & \mathbf{P}\times\mathbf{P} & \\ & & \mathbf{P}\times\mathbf{P}\times\mathbf{P} \end{pmatrix}.$$

This result may be proved in the manner of lemma 6.5.1. The generalization to fourth- and higher-order moments is obvious. In the above two-type example

$$\mathbf{p}^{(1)} = \begin{pmatrix} 0.2 \\ 0.3 \\ 0.5 \end{pmatrix}, \quad \mathbf{p}^{(2)} = \begin{pmatrix} 0.25 \\ \\ 0.75 \end{pmatrix} \quad \text{and} \quad \mathbf{P} = \begin{pmatrix} 0.2 & 0 \\ 0.3 & 0 \\ 0.5 & 0 \\ 0 & 0.25 \\ 0 & 0.75 \end{pmatrix}.$$

4. The transformation from moments about the origin to falling factorial moments (and conversely) is linear, and it is straightforward to construct the transformation matrices ($\mathbf{F}$ and $\mathbf{M}$ respectively) using the Stirling numbers $s(n, k)$ and $S(n, k)$ defined by[8]

$$(t)_n = t(t-1)(t-2)\dots(t-n+1) = \sum_{k=0}^{n} s(n,k)\, t^k;$$

$$t^n = \sum_{k=0}^{n} S(n,k)\,(t)_k.$$

If moments of first-, second- and third-order are being considered,

$$\mathbf{F} = \begin{pmatrix} \mathbf{I} & & \\ \mathbf{F}_{21} & \mathbf{I}\times\mathbf{I} & \\ \mathbf{F}_{31} & \mathbf{F}_{32} & \mathbf{I}\times\mathbf{I}\times\mathbf{I} \end{pmatrix}; \mathbf{M} = \begin{pmatrix} \mathbf{I} & & \\ \mathbf{M}_{21} & \mathbf{I}\times\mathbf{I} & \\ \mathbf{M}_{31} & \mathbf{M}_{32} & \mathbf{I}\times\mathbf{I}\times\mathbf{I} \end{pmatrix}.$$

The generalization to fourth- and higher-order moments is obvious.

Using these four results, it is soon apparent that the recurrence relation for the moments about the origin in any multi-type Galton–Watson process may be written in the linear form

$$\mathbf{m}(t+1) = \mathbf{TMBFm}(t). \tag{9.7.1}$$

$\mathbf{m}(t)$ is the vector at time t of the moments about the origin, listed in dictionary order and in increasing degree. The matrix $\mathbf{F}$ converts these moments into factorial moments and then premultiplication by the matrix $\mathbf{B}$ gives the factorial moments of the intermediary random variables. The matrix $\mathbf{M}$ then converts the factorial moments into

[8] See J. Riordan (1958), page 32, and J. H. Pollard (1969*b*).

moments about the origin, and finally premultiplication by the matrix **T** gives the moments about the origin at time $t+1$.

F and **M** are square matrices, and **B** and **T** are usually rectangular. However, **TMBF** is square, and has the form

$$\mathbf{TMBF} = \begin{pmatrix} \mathbf{A} & & & \\ \mathbf{D}_{21} & \mathbf{A}\times\mathbf{A} & & \\ \mathbf{D}_{31} & \mathbf{D}_{32} & \mathbf{A}\times\mathbf{A}\times\mathbf{A} & \\ \cdot & \cdot & \cdot & \cdot \end{pmatrix};$$

$\mathbf{A} \equiv \mathbf{QP}$ is the generalization of the usual Leslie matrix.

All the matrices **T**, **M**, **B**, and **F** have straightforward forms, and can be generated easily by a computer once the matrix **P** and the matrix **Q** are specified. These two matrices describe fully the multi-type Galton–Watson process. Five points should be noted:

1. Because the method of listing moments in the moment vector $\mathbf{m}(t)$ involves a certain amount of redundancy, the above matrix recurrence relation is *not* unique.

2. Many of the asymptotic results described in the earlier sections of this chapter may be generalized to higher-order moments about the origin.

3. From equations (9.3.12) and (9.3.13), it is known that the central quadratic moments and the quadratic moments about the origin obey the same recurrence relation. Only the initial conditions are different. This result is true for the general multi-type process, but it does not generalize to higher-order moments.

4. For a multi-type process with a large number of each type present initially, all the random variables at subsequent times are approximately distributed as correlated multivariate normal variables, due to the Central Limit Theorem, and the essential independence of the families generated by two co-existing individuals. (Some types may even possibly be absent at time $t = 0$.) In this situation, all the information required from the moments is available using the recurrence relation for expectations and central quadratic moments. This applies to most population models.

5. Precise details about the forms of the submatrices of **F** and **M** have been given by J. H. Pollard (1969*b*). The submatrices $\mathbf{F}_{21}$ and $\mathbf{M}_{21}$ are the only ones ever required in practice, and they both have simple forms. Consider first $\mathbf{F}_{21}$: the rows of this submatrix may be denoted by number pairs $(1, 1), (1, 2), \ldots, (k, k)$, and the columns by single numbers $1, 2, \ldots, k$. Then all the elements of $\mathbf{F}_{21}$ are zero except the element in the (j,j) row and the j column $(j = 1, 2, \ldots, k)$. This element is minus one. The result for $\mathbf{M}_{21}$ is strictly analogous: all its

elements are zero except the element in the (j, j) row and the j column. This element is one. The submatrices $\mathbf{F}_{21}$ and $\mathbf{M}_{21}$ appearing in the numerical examples of section 9.10 should be noted.

The results for first- and second-order moments are very useful computationally. The expectations and second-order central moments are listed in the vector $\mathbf{m}(t)$. It is *not* necessary to store the matrix $\mathbf{F}$ in the computer, since premultiplication of $\mathbf{m}(t)$ by $\mathbf{F}$ is equivalent to subtracting the expectation of each variable from the variance of the variable. It is *not* necessary to store the matrix $\mathbf{B}$, only $\mathbf{P}$, and $\mathbf{P}$ may be stored in a compact form. Premultiplication of $\mathbf{Fm}(t)$ by $\mathbf{B}$ is straightforward. It is not necessary to store $\mathbf{M}$, since premultiplication of $\mathbf{BFm}(t)$ by $\mathbf{M}$ is equivalent to the addition of the expectation of each intermediary random variable to its second-order squared factorial moment. $\mathbf{Q}$ needs to be stored, but not $\mathbf{T}$, and premultiplication by $\mathbf{T}$ is easily achieved. We then have $\mathbf{m}(t+1)$. In the population mathematics context, $\mathbf{Q}$ can usually be stored very compactly.

9.8 *Multi-type Galton–Watson processes with random branching probabilities*

One generalization of the usual multi-type Galton–Watson process is obtained by assuming that the conditional branching probabilities are themselves random variables. We shall assume that these random variables are independent of the numbers present in the population.

It is not difficult to conceive of situations in which this type of model is applicable. Consider, for example, the population model of section 9.3. It is a well-known fact that mortality rates depend upon weather conditions; a severe winter will cause the mortality rates (especially at the older ages and at the very young ages) to rise; conversely, a mild winter will mean that the mortality rates experienced are lighter than usual. Thus, there may be occasions when it is reasonable to consider the mortality probabilities as random variables.

Branching process calculations performed with fixed conditional probabilities and large populations usually lead to variances considerably smaller than those encountered in practical situations. This fact was noted at the end of section 9.6, and has been mentioned by Z. M. Sykes (1969*a*) and J. H. Pollard (1968*b*). Much of the additional variability is caused by fluctuations in the probabilities themselves.

A numerical example is instructive. Consider 1,000,000 persons subject to a mortality rate q_x, where q_x has expected value 0.002 and a standard deviation 0.0001. The variance in the number of deaths due to the finite size of the population is

$$1{,}000{,}000 \times 0.002 \times 0.998 = 1{,}996,$$

whereas the variance in the number of deaths due to random fluctuations in the mortality rate q_x is approximately

$$(1{,}000{,}000)^2 \times (0.0001)^2 = 10{,}000.$$

Thus the total variability arises from two major sources:

(i) statistical fluctuations due to the finite size of the population; and

(ii) fluctuations in the conditional probabilities themselves.

With large populations, the second source of variation is often the greater, but it is frequently neglected by mathematical demographers. The general moment recurrence relation (9.7.1) requires only minor modifications to allow for the possibility of random probabilities, which may have any joint distribution whatsoever.

Consider two random variables M and N, generally not independent, and taking non-negative integral values. Let M_1' and M_2' be random variables conditional on M, having the conditional multinomial distribution Mult $(M; P_1, P_2)$. Let N_1' be a random variable conditional on N, having the conditional multinomial (binomial) distribution Mult $(N; Q_1)$, where $Q_1 \ne 1 - P_1$. The two conditional distributions are mutually independent, and the probabilities P_1, P_2 and Q_1 are random variables, independent of M and N, having any joint distribution whatsoever. It is not difficult to prove that

$$\left.\begin{aligned} \mathscr{E}M_1' &= \mathscr{E}(P_1)\,\mathscr{E}M; \\ \mathscr{E}M_1'M_2'N_1' &= \mathscr{E}(P_1P_2Q_1)\,\mathscr{E}M(M-1)\,N. \end{aligned}\right\} \tag{9.8.1}$$

A comparison of these results with those given in section 9.7 shows that the general moment recurrence relation needs only a minor modification; it now takes the form

$$\mathbf{m}(t+1) = \mathbf{T}\mathbf{M}\mathscr{E}(\mathbf{B})\,\mathbf{F}\mathbf{m}(t). \tag{9.8.2}$$

For this type of model, $\mathscr{E}(\mathbf{P}\times\mathbf{P}) \ne \mathscr{E}(\mathbf{P})\times\mathscr{E}(\mathbf{P})$, and consequently the recurrence relation (9.8.2) applies only to moments about the origin, and *not* to central quadratic moments.

9.9 *Derivation of the discrete-time equations from D. G. Kendall's continuous-time theory**

Throughout this chapter, we have assumed the Markovian discrete-time model to be the true underlying model for the population. Some writers would regard a continuous-time model as being more appropriate and they would tend to regard a discrete-time model as an approximation to the 'true' model.[9] If this latter assumption is made, it is possible to derive the discrete-time recurrence relations from the continuous-time model and give error terms for the effects of grouping. We start with D. G. Kendall's model, described in chapter 6.

Let us define $e_{n,t}$ to be the expected number of individuals at time ta aged between na and $na+a$. Consider $e_{n+1,t+1}$ when $n \geqslant 0$. It will be assumed for simplicity that t is large, so that the initial ancestor (section 6.5) may be assumed dead. Using the notation of sections 6.4 and 6.5, we see that

$$e_{n+1,t+1} = \int_{y=na}^{na+a} {}_a p_y \, \alpha(y, ta) \, dy.$$

The function ${}_a p_y$ is well behaved, and it may be expanded in a Taylor series about the point $y = na + \frac{1}{2}a$. For typographical convenience $p, p', p'', \ldots$ will be written for ${}_a p_y$ and its y-derivatives evaluated at this point. Then

$$e_{n+1,t+1} = p\, e_{n,t} + \int_{y=na}^{na+a} \{(y-na-\tfrac{1}{2}a)\, p' + \tfrac{1}{2}(y-na-\tfrac{1}{2}a)^2 p'' + \ldots\}\, \alpha(y, ta)\, dy.$$

It is possible to expand $\alpha(y, ta)$ in a Taylor series about the point $y = na + \frac{1}{2}a$, and for typographical reasons $\alpha, \alpha', \alpha'', \ldots$ will be written for $\alpha(y, ta)$ and its y-derivatives evaluated at the point $y = na + \frac{1}{2}a$. The range of integration is symmetrical about the point $y = na + \frac{1}{2}a$, and so many of the terms integrate to zero. We find that

$$e_{n+1,t+1} = {}_a p_{na+\frac{1}{2}a}\, e_{n,t} + \tfrac{1}{12} a^3 (\alpha' p' + \tfrac{1}{2}\alpha p''). \tag{9.9.1}$$

The first term on the right-hand side of equation (9.9.1) gives the usual discrete-time survival recurrence relation, and the other term is the error term. When $\alpha(x, t)$ is the stable population (3.3.1),

$$\alpha(x, t) = A_0 e^{r_0(t-x)}\, {}_x p_0$$

and the error term in equation (9.9.1) is

$$\frac{a^2}{12} \frac{l_{x+a}}{l_x} \{(r_0 + \mu_x)(\mu_{x+a} - \mu_x) + \tfrac{1}{2}(\mu_{x+a} - \mu_x)^2 - \tfrac{1}{2}(\mu_{x+a}' - \mu_x')\}\, a\alpha(x, t),$$

[9] In many ways the best practical model would seem to be a discrete-time model with a time unit of one year. The effects of grouping are very small, and the gestation period (ignored by the continuous-time theories) causes few difficulties. The calculations are very straightforward with a computer.

evaluated at $x = na+\frac{1}{2}a$. The error term is therefore relatively small provided

$$\frac{a^2}{12}\frac{l_{x+a}}{l_x}\{(r_0+\mu_x)(\mu_{x+a}-\mu_x)+\tfrac{1}{2}(\mu_{x+a}-\mu_x)^2-\tfrac{1}{2}(\mu_{x+a}'-\mu_x')\}, \quad (9.9.2)$$

evaluated at $x = na+\frac{1}{2}a$, is small relative to unity. Numerical values of formula (9.9.2), when $a = 1$ for example, are very small (10^{-5} or less for the Australian population).

The relation for $e_{0,t+1}$ is a little more difficult to derive. It may be shown, using the techniques of section 6.5 that

$$e_{0,t+1} = \sum_{r=0}^{\infty}\int_{u=ra}^{ra+a}\alpha(u,ta)\left\{\int_{x=0}^{a} {}_{a-x}p_0\,\lambda(u+x)\,{}_xp_u\,\mathrm{d}x\right\}\mathrm{d}u.$$

The next step is to evaluate the x-integral. For typographical convenience, $\phi(x,u)$ will be written for ${}_{a-x}p_0\lambda(u+x)\,{}_xp_u$; $\phi_1(x,u)$ will denote the first partial derivative of ϕ with respect to x, and $\phi_{11}(x,u)$ the second partial derivative with respect to x. The subscript 2 will denote a partial derivative with respect to u. Using the Taylor expansion of $\phi(x,u)$ about the point $x = \frac{1}{2}a$, we find that the x-integral is equal to

$$a\phi(\tfrac{1}{2}a, u)+\tfrac{1}{12}a^3\phi_{11}(\tfrac{1}{2}a,u),$$

plus terms of smaller order. Both these terms can be expanded in power series about the point $u = ra+\frac{1}{2}a$, and so can $\alpha(u,ta)$. We then integrate with respect to u and find that

$$e_{0,t+1} = \sum_{r=0}^{\infty}\{{}_{\frac{1}{2}a}p_{ra+\frac{1}{2}a}\,a\lambda(ra+a)\,{}_{\frac{1}{2}a}p_0\}\,e_{r,t}$$

$$+\frac{a^4}{12}\sum_{r=0}^{\infty}\{\alpha'[r]\,\phi_2[r]+\alpha[r]\,\phi_{11}[r]+\tfrac{1}{2}\alpha[r]\,\phi_{22}[r]\}, \quad (9.9.3)$$

where $\alpha[r]$, $\alpha'[r]$, $\phi_2[r]$, $\phi_{11}[r]$ and $\phi_{22}[r]$ denote the various derivatives of $\alpha(u,ta)$ and $\phi(x,u)$ evaluated at the point $x = \frac{1}{2}a$, $u = ra+\frac{1}{2}a$. The first term on the right-hand side is the usual discrete-time formula, and the other term is the error term.

In order that the discrete-time deterministic model be accurate, it is necessary that the error terms in formulae (9.9.1) and (9.9.3) be small. The single initial ancestor has been neglected in the derivation of these results. There is no theoretical difficulty in including him, but the resulting formulae are then more complicated. In practical situations, there are many ancestors; provided that the initial age distribution of ancestors may be closely approximated by a smooth curve, formulae (9.9.1) and (9.9.3) may be applied approximately

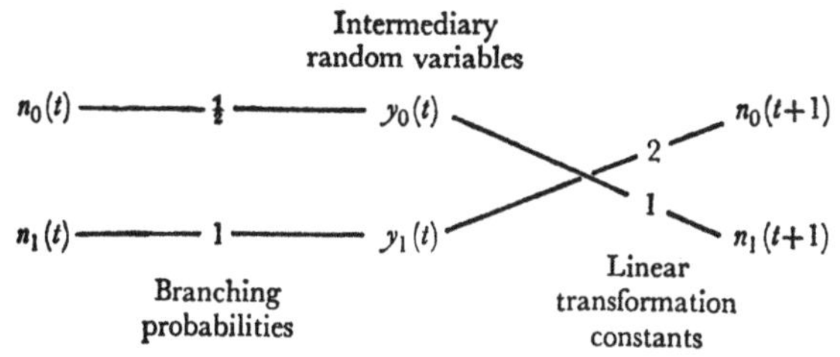

Figure 9.10.1. Diagrammatic representation of Bartlett's simple two-type process.

for all t, and not only for large t. This is what Sharpe and Lotka assumed in their original paper.

The same techniques may be applied to derive the discrete-time quadratic-moment recurrence relations of section 9.3. The algebra involved in obtaining the error terms is prohibitive, and any results obtained will be too complicated to be of any practical value. Furthermore, Kendall's model considers only single births; if the time unit is greater than one year in the discrete formulation, multiple births cannot be neglected, and the quadratic moment results derived from the continuous-time theory will be irrelevant.

9.10 *Some examples*

Consider first the two-type example in section 6.2 due to M. S. Bartlett. The process is represented diagrammatically in figure 9.10.1. For this process,

$$\mathbf{P} = \begin{pmatrix} \frac{1}{2} & 0 \\ 0 & 1 \end{pmatrix}, \quad \mathbf{Q} = \begin{pmatrix} 0 & 2 \\ 1 & 0 \end{pmatrix} \quad \text{and} \quad \mathbf{A} = \mathbf{QP} = \begin{pmatrix} 0 & 2 \\ \frac{1}{2} & 0 \end{pmatrix}.$$

The moment recurrence matrix, **TMBF**, for expectations and quadratic moments is

$$\left(\begin{array}{cc|cccc} 0 & 2 & & & & \\ 1 & 0 & & & & \\ \hline & & 0 & 0 & 0 & 4 \\ & & 0 & 0 & 2 & 0 \\ & & 0 & 2 & 0 & 0 \\ & & 1 & 0 & 0 & 0 \end{array}\right) \left(\begin{array}{cc|cccc} 1 & 0 & & & & \\ 0 & 1 & & & & \\ \hline 1 & 0 & 1 & 0 & 0 & 0 \\ 0 & 0 & 0 & 1 & 0 & 0 \\ 0 & 0 & 0 & 0 & 1 & 0 \\ 0 & 1 & 0 & 0 & 0 & 1 \end{array}\right) \left(\begin{array}{cc|cccc} \frac{1}{2} & 0 & & & & \\ 0 & 1 & & & & \\ \hline & & \frac{1}{4} & 0 & 0 & 0 \\ & & 0 & \frac{1}{2} & 0 & 0 \\ & & 0 & 0 & \frac{1}{2} & 0 \\ & & 0 & 0 & 0 & 1 \end{array}\right) \left(\begin{array}{cc|cccc} 1 & 0 & & & & \\ 0 & 1 & & & & \\ \hline -1 & 0 & 1 & 0 & 0 & 0 \\ 0 & 0 & 0 & 1 & 0 & 0 \\ 0 & 0 & 0 & 0 & 1 & 0 \\ 0 & -1 & 0 & 0 & 0 & 1 \end{array}\right).$$

The product is the moment recurrence matrix in equation (6.2.5).

The results for Bartlett's more general model (section 6.3) should be compared with the matrix equation (9.3.11). When the notation

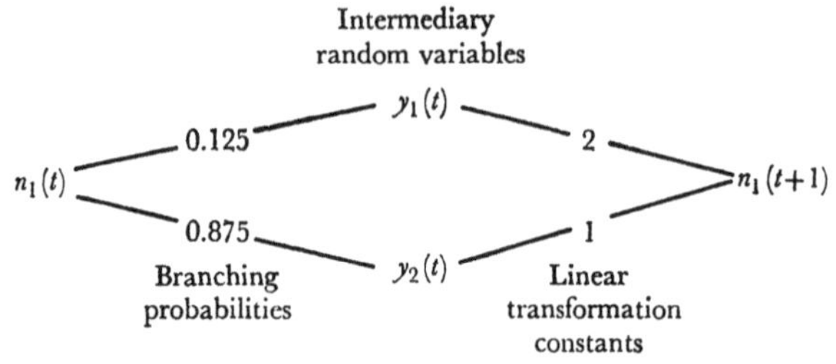

Figure 9.10.2. A simple one-type process.

is changed appropriately, and allowances are made for the fact that Bartlett ignored terms of smaller order than Δt, it is clear that the equations are equivalent.

Another numerical example is instructive; consider the process depicted in figure 9.10.2. The recurrence matrix for first- and second-order moments may be determined by hand:

$$\mathbf{TMBF} = \left(\begin{array}{cc|cccc} 2 & 1 & & & & \\ \hline & & 4 & 2 & 2 & 1 \end{array}\right) \left(\begin{array}{cc|cccc} 1 & & & & & \\ & 1 & & & & \\ \hline 1 & & 1 & & & \\ & & & 1 & & \\ & & & & 1 & \\ & 1 & & & & 1 \end{array}\right) \left(\begin{array}{c|c} 0.125 & \\ 0.875 & \\ \hline & 0.015625 \\ & 0.109375 \\ & 0.109375 \\ & 0.765625 \end{array}\right) \left(\begin{array}{c} 1 \\ \hline -1 \end{array}\right.$$

$$= \left(\begin{array}{c|c} 1.125 & \\ \hline 0.109375 & 1.265625 \end{array}\right).$$

One important practical point should be stressed. If the recurrence relation is for expectations and central quadratic moments only, and it is to be used purely for numerical calculations inside a computer, there is no need to determine the product matrix **TMBF**. The only data that must be stored in the memory are the matrices **P** and **Q**, defined in section 9.7, and certain dimensional details. It is possible to store **P** compactly, and frequently in population-model contexts, **Q** can be stored compactly as well. The calculation procedure is then very simple.

9.11 *Exercises*

1 X_1 and X_2 are two mutually independent identically distributed random variables with the uniform distribution over the unit interval (0, 1). Two random variables Y_1 and Y_2 are defined in terms of X_1 and X_2 as follows:

$$Y_1 = (-2 \log X_1)^{\frac{1}{2}} \cos (2\pi X_2);$$

and

$$Y_2 = (-2 \log X_1)^{\frac{1}{2}} \sin (2\pi X_2).$$

Prove that Y_1 and Y_2 are independent unit normal deviates (G. E. Box and M. E. Müller, 1958).

2 The natural stochastic model corresponding to P. H. Leslie's deterministic population model is a special case of the multi-type Galton–Watson process described in section 8.4. Let us imagine that there is a single initial ancestor aged zero. Prove that the probability of ultimate extinction of the population may be determined by considering a single-type process, and relate this process to the underlying multi-type process. Ignore the problem of multiple births.

3 The following matrix applies to a two-age-group population, and each element is a probability:

$$\begin{pmatrix} 0.5 & 0.75 \\ 0.9 & 0 \end{pmatrix}.$$

Use the formula developed in question 2 to determine the probability of ultimate extinction for the population which is made up initially of one female aged zero. Also use a two-type process to determine this probability.

4 What is the probability of ultimate extinction for the population in question 3 if there are initially five individuals aged zero and three aged one?

5 Determine the recurrence equation for the first- and second-order moments of the population in question 3 using the general method of section 9.7.

6 Derive equations (9.5.3).

7 Derive formulae (9.8.1).

8 Prove lemma 9.5.1.

10

Hierarchical population models and recruitment

10.1 *Introduction*

The analysis of an organization having various grades of employee has been undertaken in many different contexts. H. L. Seal, for example, in 1945 used continuous-time methods to study a population composed of k stationary strata each recruited from the stratum below and supported at the lowest level by a uniform annual number of entrants; S. Vajda in 1947 studied a discrete-time stratified population model in which all the strata were maintained at a constant level; A. Young and G. Almond considered an institution of undisclosed type having six grades of staff in 1961; J. Gani, in 1963, considered 'Australian Universities' as an organization and students were graded according to their stage of study; J. H. Pollard used Leslie-type techniques to study the size and age-structure of the Australian Academy of Science in 1964; three population models incorporating immigration were described in section 4.9. All these analyses were deterministic.

Equation (9.3.11) is important, because it suggests that whenever a Leslie deterministic method can be applied to a problem, a branching-process stochastic model can be used instead, and the expectations and quadratic moments easily calculated using a recurrence relation similar in form to the corresponding deterministic recurrence equation. We shall demonstrate the use of these techniques in this chapter.

Consider a population which survives and reproduces according to the model of section 9.3, and assume that this population is subject to immigration. Let the number of immigrants aged x entering the population at time t be a random variable $\nu_{x,t}$ with expected value $\epsilon_{x,t}$ and variance $\gamma^{(t)}_{x,x}$. The covariance $\mathrm{Cov}\,(\nu_{x,t}, \nu_{y,t})$ is denoted by $\gamma^{(t)}_{x,y}$. Furthermore, it will be assumed that the immigrants entering the population survive and reproduce in the same manner as the overall population. Then, if the numbers of immigrants can be assumed independent of the numbers in the overall population at the preceding point of time,

$$\begin{pmatrix} \mathbf{e}_{t+1} \\ \mathbf{C}(t+1) \end{pmatrix} = \begin{pmatrix} \mathbf{A} & \mathbf{0} \\ \mathbf{D} & \mathbf{A}\times\mathbf{A} \end{pmatrix} \begin{pmatrix} \mathbf{e}_t \\ \mathbf{C}(t) \end{pmatrix} + \begin{pmatrix} \boldsymbol{\epsilon}_t \\ \boldsymbol{\gamma}(t) \end{pmatrix}, \qquad (10.1.1)$$

and this equation may be written more concisely as

$$\mathbf{u}_{t+1} = \mathbf{X}\mathbf{u}_t + \boldsymbol{\varphi}_t.$$

If $\boldsymbol{\varphi}_t$ is independent of t, the subscript t can be omitted, and the equation has exactly the same form as (4.9.1). Indeed, immigration model 1 of section 4.9 is the corresponding deterministic model. It is important to note however that in general there is no one-to-one correspondence between deterministic and stochastic models.

10.2 *Gani-type models*

Consider an organization in which there are $n_{j,t}$ employees in grade j at time t ($j = 1, 2, \ldots, k$; $t = 0, 1, 2, \ldots$). During the time interval $(t, t+1)$ an employee in grade i moves to grade j with a fixed probability p_{ij}, and leaves the service with probability

$$1 - \sum_{j=1}^{k} p_{ij},$$

which is always non-zero, because there is always the possibility of death for the employee. For existing members of the organization, the process of promotion may be regarded as a multi-type Galton–Watson process, and the techniques of chapters 8 and 9 may be employed to analyse the organization structure.

In particular, the expected numbers in the various grades at time t obey a linear matrix recurrence relation of the form

$$\mathbf{e}_{t+1} = \mathbf{A}\,\mathbf{e}_t, \tag{10.2.1}$$

where

$$\mathbf{A} = \begin{pmatrix} p_{11} & p_{21} & \cdots & p_{k1} \\ p_{12} & p_{22} & \cdots & p_{k2} \\ \cdot & \cdot & \cdot & \cdot \\ p_{1k} & p_{2k} & \cdots & p_{kk} \end{pmatrix}, \tag{10.2.2}$$

and $\mathbf{e}_t$ is a k-dimensional column vector whose jth element represents the expected number of individuals in grade j at time t.

From section 9.7, we know that the expectations and central quadratic moments obey a linear recurrence relation of the form (9.3.11). This recurrence relation may be written more concisely as

$$\mathbf{m}(t+1) = \mathbf{X}\mathbf{m}(t). \tag{10.2.3}$$

Since the organization is composed of people who must leave it at some stage (if only by death)

$$\lim_{t\to\infty} \mathbf{A}^t = \mathbf{0}. \tag{10.2.4}$$

That is, the dominant latent root of $\mathbf{A}$ has modulus less than unity. (If there is an age structure in the system, and a fixed upper age limit exists, all the latent roots of $\mathbf{A}$ are zero.) The dominant latent root of $\mathbf{A} \times \mathbf{A}$ is the square of the dominant latent root of $\mathbf{A}$, and consequently[1] the dominant latent root of $\mathbf{X}$ is equal to the dominant latent root of $\mathbf{A}$, and it is less than unity in absolute value.

A random number $R(t)$ of new employees enter the organization at time t, and these go into the various grades with fixed conditional probabilities $\{a_j\}$ $(j = 1, 2, \ldots, k)$. They are assumed independent of existing members, and of existing members leaving the service.[2] Let $R(t)$ have expectation $r_1(t)$ and variance $r_2(t)$, and let these two moments be listed in a two-dimensional vector $\mathbf{r}(t)$.

The process of placing new members in the various grades is very similar to one stage of a multi-type Galton–Watson process with a non-square expectation matrix. The expectations and central quadratic moments of the new members going into the various grades are listed appropriately in a vector of the form $\mathbf{Nr}(t)$. The matrix $\mathbf{N}$ is of dimension $(k+k^2) \times 2$, and may be factored in the form

$$\mathbf{T}_1 \mathbf{M}_1 \mathbf{B}_1 \mathbf{F}_1$$

of section 9.7. In this particular case, $\mathbf{T}_1$ is of dimension

$$(k+k^2) \times (k+k^2),$$

and is really only an identity matrix; $\mathbf{M}_1$ is of dimension

$$(k+k^2) \times (k+k^2);$$

$\mathbf{B}_1$ is of dimension $(k+k^2) \times 2$; and $\mathbf{F}_1$ is of dimension 2×2.

Because of the simple additive properties of expectations and variances, the following recurrence relation is true for the process:

$$\mathbf{m}(t+1) = \mathbf{Xm}(t) + \mathbf{Nr}(t+1). \tag{10.2.5}$$

The solution depends upon the form of $\mathbf{r}(t)$. Consider, for example, the simplest case in which $\mathbf{r}(t)$ is independent of t. This case includes many possible models: for example, $R(t)$ may be Poisson with constant parameter λ, or possible $R(t)$ is a constant. Equation (10.2.5) becomes

$$\mathbf{m}(t+1) = \mathbf{Xm}(t) + \mathbf{Nr}, \tag{10.2.6}$$

with solution

$$\mathbf{m}(t) = \mathbf{X}^t \mathbf{m}(0) + (\mathbf{I} - \mathbf{X})^{-1} (\mathbf{I} - \mathbf{X}^t)\, \mathbf{Nr}. \tag{10.2.7}$$

It was noted above that the dominant latent root of $\mathbf{X}$ has modulus less than one, and so asymptotically,

$$\mathbf{m}(t) \simeq (\mathbf{I} - \mathbf{X})^{-1} \mathbf{Nr}. \tag{10.2.8}$$

A large variety of different models may be constructed and analysed along these lines.

[1] See equation (9.4.1).

[2] For this reason, the model has been called a *Gani-type model*.

10.3 *Gani-type model – numerical example*

The following matrix of probabilities was suggested by D. J. Bartholomew (1968) for an organization with five grades of staff:

$$\mathbf{A} = \begin{pmatrix} 0.65 & 0 & 0 & 0 & 0 \\ 0.20 & 0.70 & 0 & 0 & 0 \\ 0 & 0.15 & 0.75 & 0 & 0 \\ 0 & 0 & 0.15 & 0.85 & 0 \\ 0 & 0 & 0 & 0.10 & 0.95 \end{pmatrix}.$$

The submatrix $\mathbf{D}$ of the recurrence matrix for first- and second-order moments was obtained using a general computer program; and is given as equation (10.3.1).

$$\mathbf{D} = \begin{pmatrix} 0.2275 & 0 & 0 & 0 & 0 \\ -0.13 & 0 & 0 & 0 & 0 \\ 0 & 0 & 0 & 0 & 0 \\ 0 & 0 & 0 & 0 & 0 \\ 0 & 0 & 0 & 0 & 0 \\ & & & & \\ -0.13 & 0 & 0 & 0 & 0 \\ 0.16 & 0.21 & 0 & 0 & 0 \\ 0 & -0.105 & 0 & 0 & 0 \\ 0 & 0 & 0 & 0 & 0 \\ 0 & 0 & 0 & 0 & 0 \\ & & & & \\ 0 & 0 & 0 & 0 & 0 \\ 0 & -0.105 & 0 & 0 & 0 \\ 0 & 0.1275 & 0.1875 & 0 & 0 \\ 0 & 0 & -0.1125 & 0 & 0 \\ 0 & 0 & 0 & 0 & 0 \\ & & & & \\ 0 & 0 & 0 & 0 & 0 \\ 0 & 0 & 0 & 0 & 0 \\ 0 & 0 & -0.1125 & 0 & 0 \\ 0 & 0 & 0.1275 & 0.1275 & 0 \\ 0 & 0 & 0 & -0.085 & 0 \\ & & & & \\ 0 & 0 & 0 & 0 & 0 \\ 0 & 0 & 0 & 0 & 0 \\ 0 & 0 & 0 & 0 & 0 \\ 0 & 0 & 0 & -0.085 & 0 \\ 0 & 0 & 0 & 0.09 & 0.0475 \end{pmatrix}. \qquad (10.3.1)$$

There are initially 590 members of staff, and these were distributed among the five grades in the proportions 0.40, 0.30, 0.15, 0.10, and 0.05, respectively. Two types of recruitment were considered:

(*a*) $R(t)$ a fixed number which is constant over time and equal to 60; and

(*b*) $R(t)$ a Poisson random variable, independent of time, with mean equal to 60.

TABLE 10.3.1. *Expectations and covariance matrices at time $t = 100$ for the Gani-type model with fixed and with Poisson recruitment*

		Expectations at time $t = 100$		
Grade 1	Grade 2	Grade 3	Grade 4	Grade 5
128.57143	135.71429	81.42857	81.42859	162.26700
		Covariance matrix at time $t = 100$		
Fixed number of recruits				
70.12987	−34.58239	−6.57909	−1.43343	−0.24359
−34.58239	104.79128	−11.09737	−3.96612	−1.05248
−6.57909	−11.09737	74.13103	−4.34854	−1.86299
−1.43343	−3.96612	−4.34854	76.84094	−3.74366
−0.24359	−1.05248	−1.86299	−3.74366	154.49674
Poisson recruitment				
128.57143	0	0	0	0
0	135.71429	0	0	0
0	0	81.42857	0	0
0	0	0	81.42859	0
0	0	0	0	162.26015

Table 10.3.1 gives the expectations and covariance matrix in the two cases for $t = 100$. The covariance matrix for the Poisson case is of particular interest: the expectations and variances are more or less equal, and the covariances are zero to five decimal places. This suggests that the numbers in the various grades are asymptotically mutually independent Poisson variates and it is in fact true; the result is proved in section 10.4. The expectations and covariances were also output for $t = 0, 1, 2, \ldots, 99$. They are not given here however.

10.4 *The case of independent Poisson recruitment*

Let $R(t)$ be a Poisson variate with parameter $\lambda(t)$, and let us assume that the $\{R(t)\}$ are mutually independent. At time $t+T$, an individual who entered at time t may be in any one of $k+1$ categories, the extra category representing all those who have left the service. The conditional probability that such an individual is in category

$$j\,(j = 1, 2, \ldots, k+1)$$

at time $t+T$ is a multinomial probability $P_j(T)$, which is a function of the conditional probabilities $\{a_m\}$ and $\{p_{mn}\}$ and of T.

The probability that there are n_j of these characters in grade j $(j = 1, 2, \ldots, k+1)$ at time $t+T$ is equal to the conditional probability that there are n_j of these characters in grade j at time $t+T$ given that $n = \Sigma n_j$ characters entered at time t, multiplied by the probability that n characters entered at time t. That is, the probability is equal to

$$\left(\frac{n!}{n_1! \ldots n_{k+1}!} P_1^{\,n_1} \ldots P_{k+1}^{\,n_{k+1}}\right) \frac{e^{-\lambda}\lambda^n}{n!} = \prod_{j=1}^{k+1} \left\{\frac{e^{-\lambda P_j}(\lambda P_j)^{n_j}}{n_j!}\right\}.$$

Therefore the numbers of such individuals in the k grades at time $t+T$ are mutually independent Poisson variates. The total number of *recruited* individuals in grade j at time s is the sum of those in grade j at time s recruited at times $1, 2, 3, \ldots, s$. The $\{R(t)\}$ are mutually independent, and therefore the total number of *recruited* individuals in grade j at time s is the sum of s mutually independent Poisson variates, and hence a Poisson variate. Furthermore, the total numbers of recruited individuals in the k grades at a given point of time are mutually independent Poisson variates.

It was noted earlier that the dominant latent root of $\mathbf{A}$ has modulus less than unity, and this ensures that the original members become extinct with probability one (section 8.4). It follows, therefore, that the numbers in the various grades are asymptotically mutually independent Poisson variates. This result becomes important when it comes to computation, since it is only necessary to calculate expectations at time t when asymptotic results are required.

The situation may also arise in which the random number of recruits is not Poisson, but all the other conditions are true. Then, if the variances of the numbers of recruits are equal to their means, the asymptotic variances of the numbers in the k grades will be equal to their respective means, and the numbers in the k grades will be asymptotically uncorrelated.

Further, it may be deduced that if the $\{R(t)\}$ have distributions which are overdispersed (i.e. variance greater than mean), the numbers in the various grades will be asymptotically positively correlated, whereas, if the $\{R(t)\}$ are underdispersed, the numbers in the various grades are asymptotically negatively correlated. The results in table 10.3.1 exhibit this property, and so do the results in section 10.5.

10.5 *The age structures of learned societies*

The age structures of learned societies have been investigated by various authors (e.g. R. Strachey, 1892; R. Pearl, 1925; A. Schuster, 1925; A. V. Hill, 1939, 1954, 1961; and H. O. Lancaster, 1964). One point of interest in some of these analyses has been the proportion of older members (those aged 70 years or over, say). This problem is examined in this section for the Australian Academy of Science. According to the 'Report on the Population of the Academy' by H. O. Lancaster,

> 'The Academy was founded in 1954 by 23 Foundation Members who proceeded to elect 41 Ordinary Fellows. Since the original 23 were largely the Fellows of the Royal Society resident in Australia, their ages tended to have a lower limit of about 45 years. There seems to have been a conscious attempt to elect rather younger men at the first election although there were several very senior scientists and "elder statesmen" elected among the 41... The averages of the ages at the time of election for the years 1955 to 1963 were 57, 44, 47, 45, 44, 45, 46, 51 and 50. There certainly has been no tendency for the ages at election to fall since 1955.'

In table 10.5.1 the numbers of new Fellows elected in the years 1954 to 1963 are given. Table 10.5.2 gives the age distribution at time of election of Fellows elected during this period, and table 10.5.3 gives the age structure of the Academy on 1 January 1964.

Lancaster, in his report, assumes for purposes of projection that there will be exactly six new Fellows elected per annum in the future. This would seem to be a lower limit. Let us apply a Gani-type model to the problem. The main difficulty is that of finding a suitable distribution for $R(t)$, the number of new Fellows elected at time t. Two projections will therefore be made:

(*a*) Assume $R(r)$ is Poisson with mean 6. This would seem to give a lower limit to the expected number of new Fellows each year, and at the same time exaggerate the variance in the results.

(*b*) Assume $R(t)$ is Poisson with mean $\lambda(t)$, where $\lambda(t) = u + vt$, and u and v are constants. Under this assumption, the Academy will grow indefinitely, and the projection will therefore give some idea of

TABLE 10.5.1. *Number of new Fellows elected in the years* 1954 *to* 1963

Years	Number of new Fellows elected
1954	41*
1955	4
1956	5
1957	5
1958	5
1959	6
1960	6
1961	6
1962	6
1963	6

* 1954 was the first election year; these 41 Fellows are ignored in the analysis except in so far as some of them are enumerated in table 10.5.3.

TABLE 10.5.2. *Age distribution at time of election of Fellows elected during the period* 1955 *to* 1963

Age last birthday	Fellows elected 1955–63
30–4	—
35–9	6 (12.2%)
40–4	12 (24.5%)
45–9	13 (26.5%)
50–4	10 (20.4%)
55–9	2 (4.1%)
60–4	5 (10.2%)
65–9	1 (2.0%)
70–	—

the age structure when the number of Fellows elected each year is overestimated. Regression analysis applied to the data of table 10.5.1 gives us $\lambda(t) = 0.233t+6.61$, where t is the time in years measured from 1 January 1964.

The mathematical analysis of projection (*a*) is given in section 10.2. Equation (10.2.8) predicts a stable moment vector. The probability that a new Fellow elected at time t will be aged x at the time of his election was assumed to be independent of time, and estimated using the data of table 10.5.2, and a graphical method of graduation.

The equation representing projection (*b*) is

$$\mathbf{m}(t+1) = \mathbf{X}\mathbf{m}(t) + (0.233\,t+6.61)\,N\mathbf{r}, \qquad (10.5.1)$$

TABLE 10.5.3. *The number of Fellows of the Australian Academy of Science aged x last birthday on 1 January 1964*

Age	Number	Age	Number	Age	Number	Age	Number	Age	Number
35	0	45	2	55	2	65	4	75	0
36	0	46	4	56	5	66	1	76	1
37	0	47	6	57	7	67	1	77	0
38	0	48	3	58	4	68	2	78	0
39	1	49	5	59	2	69	1	79	1
40	2	50	7	60	2	70	0	80	0
41	1	51	2	61	5	71	0	81	0
42	0	52	5	62	1	72	0	82	0
43	3	53	4	63	6	73	1	83	0
44	2	54	2	64	2	74	1	84	1

TABLE 10.5.4. *Projections of the Australian Academy of Science in* A.D. 2062

	Projection (*a*)	Projection (*b*)
Expected academy size	158.82	678.36
Expected number of Fellows aged 70 years and over	39.04	147.12
Ratio of expectations	24.58%	21.69%

where $t = 0$ corresponds to 1 January 1964. This equation has an analytic solution, and we find that for large t

$$\mathbf{m}(t) \cong 0.233\, t\, (\mathbf{I} - \mathbf{X})^{-1} \mathbf{Nr}. \qquad (10.5.2)$$

That is, the same asymptotic expected age distribution is predicted as that obtained from projection (*a*). It is clear from the derivation of formula (10.5.2), however, that the approach to the stable expected age distribution will be slower in this case than with projection (*a*).

These calculations were performed using the above data and the Australian life table (males) for 1954. For $t > 60$, it is clear that all the original Fellows will have died, and consequently the number of Fellows aged 70 years and over, and the number of Fellows aged less than 70 are independent Poisson variates. The results obtained are summarized in table 10.5.4.

Because of the Poisson property (section 10.4), only expectation calculations need be carried out, and the distribution of the proportion of Fellows aged 70 years or over is readily computed. Figure 10.5.1

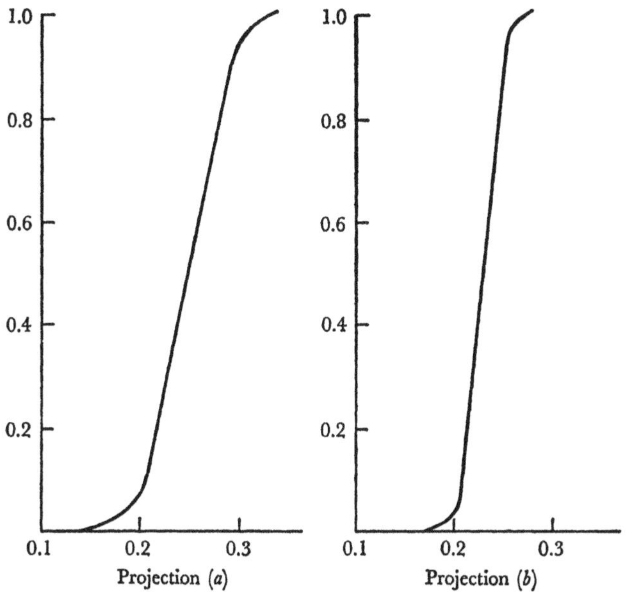

Figure 10.5.1. Continuous curves approximating the distributions of the proportions of Fellows aged 70 and over.

represents the distributions of this proportion for the two projections. The curves plotted are continuous, although of course the proportion distribution function has points of increase only at the rational points. The mean proportions are 24.58 % and 21.69% respectively.

The distributions of the proportion as $t \to \infty$ are different in the two cases. For projection (*a*), the asymptotic distribution is that given in figure 10.5.1. If Z_t denotes the proportion at time t for projection (*b*), it is soon apparent that

$$\lim_{t \to \infty} P(|Z_t - p| > \epsilon) = 0$$

for every positive number ϵ, where p is the ratio of the asymptotic expectations. That is, the proportion converges in probability to p.

The above calculations were performed assuming $R(t)$ to be Poisson; this exaggerates the computed variances, and gives numerical 'upper bounds' for the variances in the projections. In table 10.5.1, it is seen that recruitment has been virtually constant at 6 over a period of years. It is instructive, therefore, to carry out the calculations assuming $R(t)$ to be a fixed number, and equal to 6. The results are summarized in table 10.5.5. The last paragraph of section 10.4 tells us that the covariance in this table should be negative, and indeed it is.

TABLE 10.5.5. *A projection of the Australian Academy of Science in* A.D. 2062 *assuming exactly six newly-elected Fellows each year*

Expectations		
Under 70	70 and over	Expected Academy size
119.78	39.04	158.82
Covariance matrix		Variance
27.97	−9.34	40.54
−9.34	31.25	

10.6 *Young-and-Almond-type models*

Alternative models to those described in sections 10.1 to 10.5 may be obtained by assuming that $R(t)$ is made up of two independent components, the first being new members to replace those leaving or dying, and the second being a random number of new Fellows to cause the society to grow. This type of model is similar to the deterministic model constructed by A. Young and G. Almond in 1961, and described also by D. J. Bartholomew (1968). With this type of model, a member will, instead of 'dying' or 'leaving', 'move' into another category with conditional probabilities $\{a_x\}$. The same techniques may be employed, but some care is necessary, because the dominant latent root of $\mathbf{X}$ is then equal to one. (The underlying multi-type Galton–Watson process is *singular* in this case.)

10.7 *Exercises*

1 For a certain two-grade organization, there are $R(t)$ recruits at time t, where $R(t)$ is a random variable with mean $r_1(t)$ and variance $r_2(t)$. Each recruit has a chance a_1 of entering grade one and a chance $a_2 = 1 - a_1$ of entering grade two. It was shown in section 10.2 that the vector of new entrants may be written in the form $\mathbf{Nr}(t)$. Determine $\mathbf{N}$.

2 A population with a two-age-group structure has the following projection matrix:

$$\begin{pmatrix} \frac{1}{2} & \frac{2}{3} \\ \frac{3}{4} & 0 \end{pmatrix}.$$

Each element represents a probability, and births and deaths are assumed independent. What is the dominant latent root of the matrix, and what is the stable age distribution for the population?

3 A two-type stochastic process may be represented by the diagram in figure 9.7.1. Instead of multinomial distributions, the intermediary random variables $\{y_i(t)\}$ $(i = 1, \ldots, 5)$ have as conditional distributions mutually independent Poisson distributions with means $0.2n_1(t)$, $0.3n_1(t)$, $0.5n_1(t)$, $0.25n_2(t)$ and $0.75n_2(t)$. Prove that

$$\mathscr{E}y_1 = 0.2\,\mathscr{E}n_1; \qquad \mathscr{E}y_1(y_1-1) = (0.2)^2\mathscr{E}n_1^2;$$

$$\mathscr{E}y_1y_2 = (0.2)\,(0.3)\,\mathscr{E}n_1^2 \quad \text{and} \quad \mathscr{E}y_1y_4 = (0.2)\,(0.25)\,\mathscr{E}n_1n_2;$$

and deduce that the moment-recurrence equation can be written in the form

$$\mathbf{m}(t+1) = \mathbf{TMBm}(t)$$

(J. H. Pollard, 1969*b*).

4 The population in question 2 is subject to immigration. The number of immigrants at time $t+1$ is a Poisson random variable with mean equal to one half the total population size at time t, and each immigrant is of age 0 with probability $\frac{2}{3}$ and of age 1 with probability $\frac{1}{3}$. Derive a moment recurrence equation for the population, and give the moment recurrence matrix for first- and second-order moments.

5 What is the dominant latent root of the expectation recurrence matrix in question 4? Compare this latent root with that obtained in question 2.

6 What dominant latent root does the expectation recurrence matrix have when the probability that an immigrant is of age x is proportional to the population aged x in the stable population of question 2?

7 Prove the result quoted in the final paragraph of section 10.4.

8 Confirm the elements in matrix $\mathbf{D}$ of section 10.3.

9 Derive equation (10.5.2) from (10.5.1).

11

Conclusion

As this treatment of mathematical models for the growth of human populations draws to a close, some reflection on the limitations of the models is worthwhile. We do not need to dwell upon the drawbacks to the simple models which ignore age, or the short-comings of the various one-sex models, or the artificialities of some of the two-sex models; these limitations have already been discussed. Instead, we make a few general observations.

Most numerical problems in demography require mathematical models for their solution.[1] The purpose for formulating a particular model must be kept clearly in mind. It is pointless developing an extremely complex model to answer a simple question, when the validity of some of the assumptions in the model is open to dispute and the data are unreliable. Our confidence in the final answer will be no greater than our confidence in an answer obtained via a more simple model. The level of complexity of a model must be determined by the accuracy we require, the available data, and our limited understanding of the underlying social processes.

Some problems do demand rather more complicated models however. An example might be the projection of the number and age-distribution of new widows in a particular year.[2] Only a rather complex model can provide an answer, but the limitations of the data and the short-comings of the model must be remembered when the estimate is used.

We assume in all our models that the underlying rates and probabilities are independent[3] of the epoch t. This unrealistic assumption is necessary to simplify the mathematics, but quite unnecessary when practical numerical results are required. The models are readily amended to take account of secular changes[4] in the rates, the adjustments being particularly straightforward in the case of a discrete-

[1] Demographers may not always think of their calculations in this light.

[2] Such an estimate might be necessary for social security purposes.

[3] The weak-ergodicity theorem in section 4.10 is an exception.

[4] Changes over time.

time model. The recurrence equation for Leslie's model, for example, becomes

$$\mathbf{n}_{t+1} = \mathbf{A}_{t+1}\mathbf{n}_t.$$

Mathematical models are necessary for projection purposes, and models related to those in this book are normally used. Again, the complexity of the model must depend upon the purpose of the projection, the available data and our understanding of the social processes. The more detailed projection models include many refinements and it may be felt that a stochastic model, like the one in chapter 9 but also making allowance for secular changes and incorporating other features, might provide a useful estimate of variance. Small changes in the assumed trends in fertility, nuptiality and migration however have a much greater affect upon the size and structure of a projected population than do random fluctuations in the assumed rates, and we have seen in section 9.8 that the statistical variation caused by random fluctuations in the rates is much more important than the variation caused by the finite size of a population. It follows that we have little hope of obtaining a useful objective estimate of variance in a population projection. Demographers therefore often make several (deterministic) projections based on different assumptions. They can then say that *if* a certain fertility/migration pattern is followed, *then* the population will adopt a particular growth pattern; *if* another fertility/migration pattern is followed, *then* another growth pattern will be adopted.

Some of our results are remarkably robust, and we might mention the Keyfitz formula (3.8.7) as an example. The formula assumes a stable population prior to the reduction in fertility; yet it provides realistic answers with non-stable natural populations.

References

This list of references is not exhaustive, but it contains most of the references which are central to the main theme of the book. The reader is referred also to pages 155–157.

(1) Andrewartha, H. G. and Birch, L. C. (1953). 'The Lotka–Volterra theory of interspecific competition'. *Australian Journal of Zoology*, **1**, 174–7.

(2) Archbold, J. W. (1964). *Algebra*, 348 (Third edition). London: Pitman.

(3) Bartholomew, D. J. (1968). *Stochastic Models for Social Processes*, 1–70. New York: Wiley.

(4) Bartlett, M. S. (1946). *Stochastic Processes*, 36–42. (Notes of a course given at the University of North Carolina in the Fall Quarter 1946).

(5) Bartlett, M. S. (1955). *An Introduction to Stochastic Processes with Special Reference to Methods and Applications*, 40–2. Cambridge University Press.

(6) Bartlett, M. S. (1957). 'On theoretical models for competitive and predatory biological systems'. *Biometrika*, **44**, 27–42.

(7) Bartlett, M. S. and Kendall, D. G. (1951). 'On the use of the characteristic functional in the analysis of some stochastic processes occurring in physics and biology'. *Proceedings of the Cambridge Philosophical Society*, **47**, 65–76.

(8) Bartlett, M. S., Gower, J. C. and Leslie, P. H. (1960). 'A comparison of theoretical and empirical results for some stochastic population models'. *Biometrika*, **47**, 1–11.

(9) Bellman, R. A. (1960). *Introduction to Matrix Analysis*, 226–34. New York: McGraw-Hill.

(10) Bernardelli, H. (1941). 'Population waves'. *Journal of the Burma Research Society*, **31**, 1–18.

(11) Bienaymé, I. J. (1845). 'De la loi de multiplication et de la durée des familles'. *Soc. Philomath. Paris Extraits*, Series 5, 37–9.

(12) Borch, K. (1967). 'The theory of risk'. *Journal of the Royal Statistical Society*, Series B, **29**, 432–67.

(13) Box, G. E. P. and Müller, M. E. (1958). 'A note on the generation of random normal deviates'. *Annals of Mathematical Statistics*, **29**, 610–11.

(14) Bühlmann, H. (1970). *Mathematical Methods in Risk Theory*. Berlin: Springer-Verlag.

(15) Bureau of Census and Statistics, Australia. (1965). *Demography*, (Bulletin Number 83), 144.

(16) Bureau of Census and Statistics, Australia (1969). *Interim projections of the population of Australia* (1968 to 2001).

(17) Cavalli-Sforza, L. L. (1958). 'Some data on the genetic structure of human populations'. *Proceedings of the Xth International Congress on Genetics*, **1**, 389–407. New York: Pergamon, and Oxford: Clarendon Press.

(18) Coale, A. J. (1957). 'How the age distribution of a human population is determined'. *Cold Spring Harbor Symposia on Quantitative Biology* (ed. K. B. Warren), **22**, 83–9. New York: Long Island Biological Association.

(19) Cox, P. R. (1959). *Demography*, 204–35 (Third edition). Cambridge University Press.

(20) Darwin, J. H. and Williams, R. M. (1964). 'The effect of time of hunting on the size of a rabbit population'. *New Zealand Journal of Science*, **7**, 341–52.

(21) Demetrius, L. (1969). 'The sensitivity of population growth rate to perturbations in the life cycle components'. *Math. Biosci.* **4**, 129–36.

(22) de Moivre, A. (1725). *Annuities upon Lives; or the valuation of Annuities upon any number of lives; as also, of Reversions. To which is added, an appendix concerning the expectation of life and probabilities of survivorship.* London.
(23) Everett, C. J. and Ulam, S. (1948). 'Multiplicative systems in several variables III'. *Los Alamos Scientific Laboratory*, LA707.
(24) Feller, W. (1939). 'Die Grundlagen der Volterraschen Theorie des Kampfes ums Dasein in wahrscheinlichkeitstheoretischer Behandlung'. *Acta Biotheoretica*, **5**, 11–40.
(25) Feller, W. (1941). 'On the integral equation of renewal theory'. *Annals of Mathematical Statistics*, **12**, 243–67.
(26) Feller, W. (1957). *An Introduction to Probability Theory and Its Applications*, 267, 397–421 (Second edition). New York: Wiley.
(27) Fisher, R. A. (1922). 'On the dominance ratio'. *Proceedings of the Royal Society of Edinburgh*, **42**, 321–41.
(28) Fisher, R. A. (1930). *The Genetical Theory of Natural Selection.* Oxford: Clarendon Press.
(29) Freeman, H. (1962). *Finite Differences for Actuarial Students*, 1–25, 126–7, 188–9. Cambridge University Press.
(30) Frobenius, G. (1912). 'Uber Matrizen aus nicht negativen Elementen'. *Sitzungsberichte der Kgl. Preussischen Akademie der Wissenschaften*, 456–77. Berlin.
(31) Galton, F. (1873). 'Problem 4001'. *Educational Times* (1 April 1873), 17.
(32) Galton, F. and Watson, H. W. (1873). 'Problem 4001'. *Educational Times* (1 August 1873), 115.
(33) Gani, J. (1963). 'Formulae for projecting enrolments and degrees awarded in universities'. *Journal of the Royal Statistical Society*, Series A, **126**, 400–9.
(34) Gantmacher, F. R. (1959). *The Theory of Matrices.* New York: Chelsea Publishing Company.
(35) Gompertz, B. (1825). 'On the nature of the function expressive of the law of human mortality and on a new mode of determining the value of life contingencies'. *Philosophical Transactions of the Royal Society*, **115**, 513–85.
(36) Goodman, L. A. (1953). 'Population growth of the sexes'. *Biometrics*, **9**, 212–25.
(37) Goodman, L. A. (1967*a*). 'On the reconciliation of mathematical theories of population growth'. *Journal of the Royal Statistical Society*, Series A, 541–53.
(38) Goodman, L. A. (1967*b*). 'The probabilities of extinction for birth and death processes that are age-dependent or phase-dependent'. *Biometrika*, **54**, 579–96.
(39) Goodman, L. A. (1967*c*). 'On the age-sex composition of the population that would result from given fertility and mortality conditions'. *Demography*, **4**, 423–41.
(40) Goodman, L. A. (1968). 'An elementary approach to the population projection-matrix and to the mathematical theory of population growth'. *Demography*, **5**, 382–409.
(41) Goodman, L. A. (1969). 'The analysis of population growth when the birth and death rates depend upon several factors'. *Biometrics*, **25**, 659–81.
(42) Goodman, L. A. (1971). 'On the sensitivity of the intrinsic growth rate to changes in the age-specific birth and death rates'. *Theoretical Population Biology*, **2**, 339–54.
(43) Goubert, P. (1960). *Beauvais et les Beauvaisis de 1600 à 1730*; contribution à l'histoire sociale de la France du XVIIe siècle. Paris: SEVPEN.
(44) Graunt, J. (1662). *Natural and Political Observations Mentioned in a Following Index, and Made upon the Bills of Mortality, with Reference to the Government, Religion, Trade, Growth, Air, Diseases, and the Several Changes of the Said City.* London: John Martyn.
(45) Hadwiger, H. (1940). 'Eine analytische Reproduktionsfunktion für biologische Gesamtheiten'. *Skandinavisk Aktuarietidskrift*, **23**, 101–13.
(46) Hajnal, J. (1948). 'Some comments on Mr Karmel's paper: "The relations between male and female reproduction rates"'. *Population Studies*, **2**, 354–60.
(47) Hajnal, J. (1955). 'The prospects for population forecasts'. *Journal of the American Statistical Association*, **50**, 309–22.
(48) Harris, T. E. (1951). 'Some mathematical models for branching processes'. *Second Berkeley Symposium*, 305–28.
(49) Harris, T. E. (1963). *The Theory of Branching Processes*, 1–49. Berlin: Springer-Verlag.

(50) Hartree, D. R. (1958). *Numerical Analysis*, 33–46, 63, 98–101 (Second edition). Oxford: Clarendon Press.
(51) Hawkins, D. and Ulam, S. (1944). 'Theory of multiplicative processes, I'. *Los Alamos Scientific Laboratory*, LADC-265.
(52) Heyde, C. C. and Seneta, E. (1972). 'The simple branching process, a turning point test and a fundamental inequality: a historical note on I. J. Bienaymé'. *Biometrika*, **59**.
(53) Hill, A. V. (1939). 'Age of election to the Royal Society'. *Notes and Records of the Royal Society*, **2**, 71–3.
(54) Hill, A. V. (1954). 'Age of election to the Royal Society'. *Notes and Records of the Royal Society*, **11**, 14–16.
(55) Hill, A. V. (1961). 'Age of election to the Royal Society'. *Notes and Records of the Royal Society*, **16**, 151–3.
(56) Hooker, P. F. and Longley-Cook, L. H. (1953). *Life and Other Contingencies* (Two volumes). Cambridge University Press.
(57) Jordan, C. W. (1967). *Life Contingencies*, 1–170, 271–90 (Second edition). Chicago: Society of Actuaries.
(58) Karmel, P. H. (1947). 'The relation between male and female reproduction rates'. *Population Studies*, **1**, 249–74.
(59) Kendall, D. G. (1948*a*). 'On some modes of population growth leading to R. A. Fisher's logarithmic series distribution'. *Biometrika*, **35**, 6–15.
(60) Kendall, D. G. (1948*b*). 'On the role of variable generation time in the development of a stochastic birth process'. *Biometrika*, **35**, 316–39.
(61) Kendall, D. G. (1948*c*). 'On the generalized "birth-and-death" process'. *Annals of Mathematical Statistics*, **19**, 1–15.
(62) Kendall, D. G. (1949). 'Stochastic processes and population growth'. *Journal of the Royal Statistical Society*, Series B, **11**, 230–64.
(63) Kendall, D. G. (1952). 'Les processes stochastiques de croissance en biologie'. *Annales de l'Institut Henri Poincaré*, **13**, 43–108.
(64) Kendall, D. G. (1960). 'Birth and death processes, and the theory of carcinogenesis'. *Biometrika*, **47**, 13–21.
(65) Kendall, D. G. (1966). 'Branching processes since 1873'. *Journal of the London Mathematical Society*, **41**, 385–406.
(66) Kendall, M. G. and Stuart, A. (1963). *The Advanced Theory of Statistics* (Three volumes), Volume 1, 67–89 (Second edition). London: Griffin.
(67) Keyfitz, N. (1964). 'The intrinsic rate of natural increase and the dominant latent root of the projection matrix'. *Population Studies*, **18**, 293–308.
(68) Keyfitz, N. (1967). 'Reconciliation of population models: matrix, integral equation, and partial fraction'. *Journal of the Royal Statistical Society*, Series A, **130**, 61–83.
(69) Keyfitz, N. (1968). *Introduction to the Mathematics of Population*, 271–92. Reading, Mass.: Addison-Wesley.
(70) Keyfitz, N. (1971). 'On the momentum of population growth'. *Demography*, **8**, 71–80.
(71) Kuczynski, R. R. (1932). *Fertility and Reproduction*, 36–8. New York: Falcon Press.
(72) Lancaster, H. O. (1964). 'Report on the population of the Academy'. (An unpublished report on the composition of the Australian Academy of Science.)
(73) Le Bras, M. H. (1969). 'Retour d'une population à l'état stable après une "catastrophe"'. *Population*, **24**, 861–96.
(74) Lefkovitch, L. P. (1964). 'The growth of restricted populations of *Lasioderma Serricorne* (F.) (*Coleoptera, Anobiidae*)'. *Bulletin of Entomological Research*, **55**, 87–96.
(75) Lefkovitch, L. P. (1965). 'The study of population growth in organisms grouped by stages'. *Biometrics*, **21**, 1–18.
(76) Lefkovitch, L. P. (1966*a*). 'The effects of adult emigration on populations of *Lasioderma Serricorne* (F.)'. *Oikos*, **15**, 200–10.
(77) Lefkovitch, L. P. (1966*b*). 'A population growth model incorporating delayed responses'. *Bulletin of Mathematical Biophysics*, **28**, 219–33.

(78) Leslie, P. H. (1945). 'On the use of matrices in certain population mathematics'. *Biometrika*, **33**, 183–212.
(79) Leslie, P. H. (1948). 'Some further notes on the use of matrices in population mathematics'. *Biometrika*, **35**, 213–45.
(80) Lewis, E. G. (1942). 'On the generation and growth of a population'. *Sankhyā*, **6**, 93–6.
(81) Lopez, A. (1961). *Problems in Stable Population Theory*, 47–62. Princeton: Office of Population Research.
(82) Lotka, A. J. (1907). 'Mode of growth of material aggregates'. *American Journal of Science*, **24**, 199–216.
(83) Lotka, A. J. (1931 *a*). 'Population analysis – the extinction of families – I'. *Journal of the Washington Academy of Sciences*, **21**, 377–80.
(84) Lotka, A. J. (1931 *b*). 'Population analysis – the extinction of families – II'. *Journal of the Washington Academy of Sciences*, **21**, 453–59.
(85) Lotka, A. J. (1932). 'The growth of mixed populations: two species competing for a common food supply'. *Journal of the Washington Academy of Sciences*, **21**, 461–9.
(86) MacArthur, R. (1968). 'Selection for life tables in periodic environments'. *American Naturalist*, **102**, 381–3.
(87) McFarland, D. D. (1969). 'On the theory of stable population; a new and elementary proof of the theorems under weaker assumption'. *Demography*, **6**, 301–22.
(88) McKendrick, A. G. (1926). 'Applications of mathematics to medical problems'. *Proceedings of the Edinburgh Mathematical Society*, **40**, 98–130.
(89) Makeham, W. M. (1860). 'On the law of mortality'. *Journal of the Institute of Actuaries*, **13**, 325–58.
(90) Malthus, T. R. (1798). *An Essay on the Principle of Population*, Chapter 2, p. 21. (First edition). Printed for J. Johnson in St Paul's Churchyard, London.
(91) Moran, P. A. P. (1962). *Statistical Processes of Evolutionary Theory*, 1–20. Oxford: Clarendon Press.
(92) Parzen, E. (1960). *Modern Probability Theory and its Applications*, 416. New York: Wiley.
(93) Pearl, R. (1925). 'Vital statistics of the National Academy of Sciences'. *Proceedings of the National Academy of Sciences of the U.S.A.*, **11**, 752–68.
(94) Perron, O. (1907). 'Zur Theorie der Matrizen'. *Mathematische Annalen*, **64**, 248–63.
(95) Pollard, A. H. (1948). 'The measurement of reproductivity'. *Journal of the Institute of Actuaries*, **74**, 288–318.
(96) Pollard, A. H. (1949). 'Methods of forecasting mortality using Australian data'. *Journal of the Institute of Actuaries*, **75**, 151–82.
(97) Pollard, G. N. (1969). 'Factors affecting the sex ratio at birth in Australia, 1902–65'. *Journal of Biosocial Science*, **1**, 125–44.
(98) Pollard, J. H. (1966). 'On the use of the direct matrix product in analysing certain stochastic population models'. *Biometrika*, **53**, 397–415.
(99) Pollard, J. H. (1967). 'Hierarchical population models with Poisson recruitment'. *Journal of Applied Probability*, **4**, 209–13.
(100) Pollard, J. H. (1968 *a*). 'The multi-type Galton–Watson process in a genetical context'. *Biometrics*, **24**, 147–58.
(101) Pollard, J. H. (1968 *b*). 'A note on multi-type Galton–Watson processes with random branching probabilities'. *Biometrika*, **55**, 589–90.
(102) Pollard, J. H. (1968*c*). 'A note on the age structures of learned societies'. *Journal of the Royal Statistical Society*, Series A, **131**, 569–78.
(103) Pollard, J. H. (1969 *a*). 'Continuous-time and discrete-time models of population growth'. *Journal of the Royal Statistical Society*, Series A, **132**, 80–8.
(104) Pollard, J. H. (1969 *b*). 'A discrete-time, two-sex, age-specific, stochastic population program incorporating marriage'. *Demography*, **6**, 185–221.
(105) Pollard, J. H. (1970 *a*). 'A Taylor series expansion for Lotka's *r*'. *Demography*, **7**, 151–4.
(106) Pollard, J. H. (1970 *b*). 'On simple approximate calculations appropriate to populations with random growth rates'. *Theoretical Population Biology*, **1**, 208–18.

(107) Pollard, J. H. (1971). 'Mathematical models of marriage'. Paper presented at the Fourth Conference on the Mathematics of Population held in Hawaii.

(108) Rhodes, E. C. (1940). 'Population mathematics'. *Journal of the Royal Statistical Society*, **103**, 61–89, 218–45, 362–87.

(109) Riordan, J. (1958). *An Introduction to Combinatorial Analysis*, 32. New York: Wiley.

(110) Schuster, A. (1925). 'On the life statistics of Fellows of the Royal Society'. *Proceedings of the Royal Society*, **107**, A, 368–76.

(111) Seal, H. L. (1945). 'The mathematics of a population composed of k stationary strata each recruited from the stratum below and supported at the lowest level by a uniform annual number of entrants'. *Biometrika*, **33**, 226–30.

(112) Sharpe, F. R. and Lotka, A. J. (1911). 'A problem in age distribution'. *Philosophical Magazine*, **21**, 435–8.

(113) Sneddon, I. N. (1957). *Elements of Partial Differential Equations*, 49–55. New York: McGraw-Hill.

(114) Steffensen, J. F. (1930). 'Om sandsynligheden for at afkommet uddør'. *Matematisk Tidsskrift*, B, **1**, 19–23.

(115) Strachey, R. (1892). 'On the probable effect of the limitation of the number of Ordinary Fellows elected into the Royal Society to fifteen in each year on the eventual total number of Fellows'. *Proceedings of the Royal Society*, **51**, 463–70.

(116) Sutherland, I. (1963). 'John Graunt – a tercentenary tribute'. *Journal of the Royal Statistical Society*, Series A, **126**, 537–56.

(117) Sykes, Z. M. (1969*a*). 'Some stochastic versions of the matrix model for population dynamics'. *Journal of the American Statistical Association*, **64**, 111–30.

(118) Sykes, Z. M. (1969*b*). 'On discrete stable population theory'. *Biometrics*, **25**, 285–93.

(119) Tallis, G. M. (1966). 'A migration model'. *Biometrics*, **22**, 409–12.

(120) Turnbull, H. W. and Aitken, A. C. (1951). *An Introduction to the Theory of Canonical Matrices*, 58–76. New York: Dover.

(121) Usher, M. B. (1966). 'A matrix approach to the management of renewable resources, with special reference to selection forests'. *Journal of Applied Ecology*, **3**, 355–67.

(122) Usher, M. B. (1969*a*). 'A matrix model for forest management'. *Biometrics*, **25**, 309–15.

(123) Usher, M. B. (1969*b*). 'A matrix approach to the management of renewable resources, with special reference to selection forests – two extensions'. *Journal of Applied Ecology*, **6**, 347–8.

(124) Vajda, S. (1947). 'The stratified semistationary population'. *Biometrika*, **34**, 243–54.

(125) Watson, H. W. and Galton, F. (1874). 'On the probability of extinction of families'. *Journal of the Anthropological Institute of Great Britain and Ireland*, **4**, 138–44.

(126) Waugh, W. A. O'N. (1955). 'An age-dependent birth and death process'. *Biometrika*, **42**, 291–306.

(127) Waugh, W. A. O'N. (1961). 'Age dependence in a stochastic model of carcinogenesis'. *Fourth Berkeley Symposium*, **4**, 405–13.

(128) Weiler, H. (1959). 'Sex ratio and birth control'. *American Journal of Sociology*, **65**, 298–9.

(129) Wicksell, S. D. (1931). 'Nuptiality, fertility and reproductivity'. *Skandinavisk Aktuarietidskrift*, **14**, 125–57.

(130) Wrigley, E. A. (1966) (editor). *An Introduction to English Historical Demography from the Sixteenth to the Nineteenth Century*. Cambridge Group for the History of Population and Social Structure, Publication Number 1. London: Weidenfeld and Nicolson.

(131) Wrigley, E. A. and Schofield, R. S. (1968). 'A social survey of the past'. *Social Science Research Council Newsletter*, Number 2, February 1968, 16–18.

(132) Yntema, L. (1952). *Mathematical Models of Demographic Analysis*. Leiden: J. J. Groen and Zoon.

(133) Young, A. and Almond, G. (1961). 'Predicting distributions of staff'. *Computer Journal*, **3**, 246–50.

(134) Yule, G. U. (1924). 'A mathematical theory of evolution based on the conclusions of Dr J. C. Willis, F.R.S.'. *Philosophical Transactions of the Royal Society*, B, **213**, 21–87.

References according to chapter and section

Some of the references are not mentioned in the text, but they provide a convenient starting point for further reading.

Chapter 1 *Introduction*
(17), (43), (44), (90), (116), (130), (131).

Chapter 2 *The life table*

2.1 (15), (44), (56), (57), (116).
2.2 (56), (57).
2.3 (56,) (57).
2.4 (29), (50), (56), (57).
2.5 (56), (57).
2.6 (56), (57).
2.7 (29), (50).
2.8 (56), (57).
2.9 (56), (57).
2.10 (56), (57).
2.11 (56), (57).
2.12 (56), (57).
2.13 (22), (35), (56), (57), (89).
2.14 (12), (14), (29), (35), (50), (56), (57), (89).

Chapter 3 *The deterministic population models of T. Malthus, A. J. Lotka, and F. R. Sharpe and A. J. Lotka*

3.1 (82), (90), (106), (108), (112).
3.2 (25), (82), (90), (108), (112).
3.3 (108), (112).
3.4 (66), (73), (105), (108).
3.5 (69), (73), (105).
3.6 (25).
3.7 (45), (129).
3.8 (70).
3.9 —
3.10 (69), (105), (108).

Chapter 4 *The deterministic theory of H. Bernardelli, P. H. Leslie and E. G. Lewis*

4.1 (10), (40), (67), (78), (80), (91), (98).
4.2 (78), (91).
4.3 (30), (34), (49), (94).
4.4 (28), (78), (91).
4.5 (78), (91), (120).
4.6 (78).
4.7 (21), (42).
4.8 (79).

4·9 (20), (33), (74), (75), (76), (77), (102), (119), (121), (122), (123).
4·10 (18), (81), (87).
4·11 (1), (6), (8), (69), (85).
4·12 (86).
4·13 (10), (20).

Chapter 5 *Simple birth and death processes*

5·1 (24), (26), (60), (61), (62), (63), (64), (88), (126), (127), (134).
5·2 (26).
5·3 (26), (134).
5·4 (24), (26), (113).
5·5 (59), (62), (113).
5·6 (26), (59), (62).
5·7 (26), (59), (62), (113).

Chapter 6 *The stochastic models of M. S. Bartlett and D. G. Kendall*

6.1 (4), (62).
6.2 (4), (66), (98).
6.3 (4).
6.4 (4), (7), (62).
6.5 (62).
6.6 (4), (62).

Chapter 7 *The two-sex problem*

7·1 (46), (58), (71), (95).
7·2 (95), (132).
7·3 (62).
7·4 (62), (107).
7·5 (107).
7·6 (36), (58).
7·7 (39), (41).
7·8 (97), (128).
7·9 (95), (132).

Chapter 8 *The extinction of surnames*

8.1 (11), (27), (28), (31), (32), (49), (52), (65), (114), (125).
8.2 (49).
8.3 (83), (84).
8.4 (23), (48), (49), (51).
8.5 (5), (27), (49), (100).
8.6 (49).

Chapter 9 *The stochastic version of Leslie's model*

9·1 (91), (98), (118).
9·2 (2), (9), (98), (120).
9·3 (49), (98).
9·4 (23), (48), (92), (98).
9·5 (98).
9·6 (13), (98).
9·7 (98), (104), (109).
9·8 (101), (117), (118).
9·9 (37), (62), (68), (103).
9·10 (4), (98).
9·11 (13), (38), (98).

Chapter 10 ***Hierarchical population models and recruitment***

10.1 (3), (20), (33), (74), (75), (76), (77), (111), (119), (124), (133).
10.2 (3), (33), (99).
10.3 (3), (33), (99).
10.4 (3), (99).
10.5 (53), (54), (55), (72), (93), (102), (110), (115).
10.6 (3), (133).
10.7 (3).

Chapter 11 ***Conclusion***

(16), (19), (47), (78), (96).

Solutions to exercises

Chapter 2

1 (*a*) 0.980; (*b*) 0.0660; (*c*) 0.00681; (*d*) 0.0463.

3 (i) Prove that $\log p_x \doteqdot -\frac{1}{2}(\mu_x+\mu_{x+1})$. Then $q_{50} = 0.01130$.
(ii) 0.00547.
(iii) 0.00547.

4 It is possible for μ_x to be greater than one, because μ_x is *not* a probability. The functions p_x and q_x on the other hand are probabilities, and they therefore lie in the range (0, 1). It is easy to prove that

$$p_x = \exp\left(-\int_0^1 \mu_{x+t}\mathrm{d}t\right).$$

It is clear that p_x and q_x lie in the unit interval (0, 1) provided $\mu_{x+t} \geqslant 0$.

5 (i) 110;

(ii) $\mu_x = (80+2x)/(20{,}900-80x-x^2)$;

$q_x = (81+2x)/(20{,}900-80x-x^2)$;

(iii) ${}_{10}p_{20} = 0.931$.

6 $l_x = K\{(b_0+b_1x)/(a_0+a_1x)\}^n$, where $n = (a_1b_0-a_0b_1)^{-1}$.

7 $s = \mathrm{e}^{-A}$ and $g = \exp\{-B/(\log c)\}$.

8 (i) l_x must be a finite non-negative, monotonic non-increasing function of x over the age-range of the individuals. For all *real* populations it must be monotonic strictly decreasing.
(ii) Clearly $l_x = 1-x/106$ satisfies these requirements for $0 \leqslant x \leqslant 106$.
(iii) 106; (iv) $(106-x)^{-1}$; (v) $1-63/93 = 0.323$.

9 (*a*) Use formula (2.8.1) to prove that $e_x/(1+e_{x+1}) = p_x$. Then

$p_xp_{x+1}\cdots p_{x+n-1} = {}_np_x$.

(*b*) Assume a uniform distribution of deaths over each year of age. Then

${}_{\frac{1}{2}}p_x - {}_1p_x = d_x/(2l_x)$.

Recall that

$$\sum_{t=0}^{\infty} d_{x+t} = l_x,$$

and result follows.

10 See section 2.8.

11 One can use a conditional probability argument:

$$\Pr(A|B) = \Pr(A \cap B)/\Pr(B).$$

Consider a person aged 0 (although it could be any age less than x). The probability that he dies between ages x and y is $(l_x - l_y)/l_0$, where $y > x$. This is Pr (B). The probability that he dies at age $x+t$ between x and y is $\{(l_{x+t}\mu_{x+t})/l_0\}\,dt$, where $t \leqslant y-x$. This is Pr $(A \cap B)$. It follows that the conditional average age at death is

$$\int_0^{y-x} (x+t)\frac{l_{x+t}\mu_{x+t}}{l_x - l_y}\,dt,$$

which reduces to the answer.

12 (i) $l_0 e^{-\mu x}\mu\, dx$; (ii) $l_x = l_0 e^{-\mu x}$;
(iii) $1/\mu$; (iv) $\mathring{e}_x = 1/\mu$.

13 $\frac{\partial}{\partial t}\,{}_tp_x = -{}_tp_x\,\mu_{x+t}$; $\frac{\partial}{\partial x}\,{}_tp_x = -{}_tp_x(\mu_{x+t} - \mu_x)$.

Numerical values: -0.0294 and -0.0111 respectively.

14 The probability that (x) dies first $= \int_0^{\infty} {}_tp_x\,{}_tp_y\,\mu_{x+t}\,dt$.

According to example 2 of section (2.13), for Gompertz mortality

$${}_tp_x = g^{c^x(c^t-1)} = \exp\{\theta c^x(c^t - 1)\}, \text{ say.}$$

The integral becomes rather complicated, but substitute $u = c^t$. The answer is $c^x/(c^x + c^y)$.

15 The probability that (x_1) dies first of n lives $(x_1), \ldots (x_n)$ is equal to

$$c^{x_1}\Big/\sum_{j=1}^{n} c^{x_j}.$$

16 $D = \log(1+\Delta)$
$D\mathring{e}_x = (\Delta - \frac{1}{2}\Delta^2 + \frac{1}{3}\Delta^3 - \ldots)\,\mathring{e}_x$ (numerical differentiation).
We find that $\mu_{60} = 0.0219$ compared with the tabulated value 0.02094.

17 (i) $\mu_x = \{2(121-x)\}^{-1}$. (ii) $q_x = 1 - \{(120-x)/(121-x)\}^{\frac{1}{2}}$.
(iii) Probability $= (\sqrt{100} - \sqrt{81})/\sqrt{121} = 1/11$.

18 $\frac{d}{dx}L_x = \frac{d}{dx}\int_0^1 l_{x+t}\,dt = \int_0^1 \frac{d}{dx}(l_{x+t})\,dt = \int_0^1 \frac{d}{dt}(l_{x+t})\,dt = -d_x.$

This gives the first result. To prove the second, note that

$$L_x \doteqdot l_{x+\frac{1}{2}}.$$

19 Probability $= \frac{1}{l_{65}l_{75}}\int_0^{20} l_{65+t}\,l_{75+t}\,\mu_{75+t}\,dt.$

The repeated Simpson rule is

$$\int_0^{2nh} U_x \mathrm{d}x = \tfrac{1}{3}h(U_0+4U_1+2U_2+4U_3+\ldots+4U_{2n-1}+U_{2n}).$$

t	l_{65+t}	l_{75+t}	μ_{75+t}	Simpson coeff.	Product
0	67,699	39,984	0.07804	1	2.113×10^8
5	54,944	24,669	0.11823	4	6.409×10^8
10	39,984	11,758	0.18272	2	1.718×10^8
15	24,669	3,800	0.27312	4	1.024×10^8
20	11,758	742	0.38322	1	0.033×10^8
					11.297×10^8

So the probability $= \dfrac{11.297\times 10^8\times 5}{3\times 67{,}699\times 39{,}984} = 0.70.$

20 $x = \tfrac{1}{2}(x_1+x_2)$.

21 Prove first that $p_x' = p_x^2$, so that $q_x' = q_x(1+p_x)$. Hence q_x' is *less* than double q_x.

22 $\log(l_1/l_0) = -\int_0^{\frac{1}{2}}(0.15-0.10x)\,\mathrm{d}x - \int_{\frac{1}{2}}^{1}(0.01)^x\mathrm{d}x$. So $l_1 \doteqdot 92{,}127$.

23 Construct a difference table for the number of deaths between exact age 72 and age 72 plus j months.

j	Deaths	Δ	Δ^2	Δ^3	Δ^4
0	0				
		11			
1	11		5		
		16		1	
2	27		6		1
		22		2	
3	49		8		1
		30		3	
4	79		11		
		41			
5	120				

Fourth differences are constant.

The number of deaths between exact age 72 and exact age $72+t$ is

$$f(t) = \int_0^t l_{72+u}\,\mu_{72+u}\,\mathrm{d}u \quad (0 \leqslant t < 1),$$

and $l_{72} = 21{,}500$. But from the difference table we see that $f(t)$ is a fourth degree function of t, and we find that

$$f(t) = 864\,t^4 - 144\,t^3 + 354\,t^2 + 103\,t.$$

(i) We need to evaluate $L_{72} = \int_0^1 l_{72+t}\,dt$

$$= \int_0^1 \frac{d}{dt}(t-1)\, l_{72+t}\,dt$$

$$= l_{72} - \int_0^1 (1-t)\, l_{72+t}\mu_{72+t}\,dt$$

$$= l_{72} - \int_0^1 (1-t) f'(t)\,dt$$

$$= 21{,}193.7.$$

(ii) $d_{72} = f(1) = 1{,}177$

$q_{72} = 1{,}177/21{,}500 = 0.0547.$

(iii) $m_{72} = 1{,}177/21{,}193.7 = 0.0555.$

(iv) $l_{72.5}\mu_{72.5} = f'(\tfrac{1}{2}) = 781.$

$l_{72.5} = l_{72} - f(\tfrac{1}{2}) = 21{,}324.$

So $\mu_{72.5} = 781/21{,}324 = 0.0366.$

24 (i) $\int_{t=0}^1 P_x(t)\,dt.$ (ii) $\theta_x\left\{\int_0^1 P_x(t)\,dt\right\}^{-1}$ estimates m_x.

(iii) m_x is estimated by $\theta_x/P_x(\tfrac{1}{2})$.

$$m_{41} = \frac{25}{3{,}102} = 0.00806 \quad \text{for example.}$$

To evaluate q_x, we make use of formula (2.9.3), and L_x is evaluated via formula (2.9.1).

Age x	l_x	d_x	p_x	q_x	L_x	m_x
41	10,000	80	0.99197	0.00803	9,960	0.00806
42	9,920	81	0.99183	0.00817	9,880	0.00820
43	9,839	82	0.99163	0.00837	9,798	0.00841
44	9,757	82	0.99156	0.00844	9,716	0.00848
45	9,675	83	0.99138	0.00862	9,634	0.00866

This question helps explain the importance of m_x. In practice, the crude m_x values are smoothed (or graduated) before the other life table functions are computed.

25 $\bar{A}_x = \int_0^\infty v^t\, {}_tp_x\mu_{x+t}\,dt.$

26 Variance $= \int_0^\infty v^{2t}\, {}_tp_x\mu_{x+t}\,dt - (\bar{A}_x)^2.$

27 Substituting the second of formulae (2.12.9) into the first, and replacing q_x^α in the right-hand side by $(aq)_x^\alpha$, we have

$$q_x^\alpha = (aq)_x^\alpha\{1-\tfrac{1}{2}(aq)_x^\alpha\}\{1-\tfrac{1}{2}(aq)_x^\alpha - \tfrac{1}{2}(aq)_x^\beta\}^{-1}.$$

This formula may be used as an alternative to iteration or the solution of a quadratic equation. The force of decrement μ_x^α can be calculated using formula (2.13.1):

$$\mu_x^\alpha \doteqdot -\tfrac{1}{2}(\log p_{x-1}^\alpha + \log p_x^\alpha).$$

28 $_tp_x^\beta = {}_t(ap)_x / {}_tp_x^\alpha = \{1 - t(aq)_x\}\{1 - tq_x^\alpha\}^{-1},$

$(aq)_x = q_x^\alpha + q_x^\beta - q_x^\alpha q_x^\beta.$

Expand the denominator by the negative binomial theorem, and the first formula is obtained. The second formula is derived from the equation

$$(aq)_x^\alpha = \int_0^1 {}_tp_x^\beta\, {}_tp_x^\alpha \mu_{x+t}^\alpha\, dt$$

using the first result and equation (2.12.6).

29 Ignoring the possibility of theft, the probability that a light-bulb will expire between ages x and $x + dx$ is

$$\frac{l_x^d}{l_0^d}\mu_x^d\, dx = (a + 2bx)\exp\{-(bx^2 + ax)\}\, dx.$$

Now

$$l_x^d = \int_x^\infty l_t^d \mu_t^d\, dt = l_0^d \exp\{-(bx^2 + ax)\},$$

and

$$\mu_x^d = -\frac{d}{dx}\log l_x^d = a + 2bx.$$

(i) The total force of decrement $(a\mu)_x = a + \tau + 2bx$.

(ii) The life-time probability-density function is

$$(a + \tau + 2bx)\exp\{-(bx^2 + ax + \tau x)\}.$$

(iii) $_t(ap)_x = \exp\{-(bt^2 + 2btx + at + \tau t)\}$.

30 If $(al)_x^\alpha$ can be approximated by a polynomial of degree 4, the formula can be deduced from equations (2.4.1) and (2.11.13).

Chapter 3

1 The use of equation (3.4.5) is equivalent to the assumption of a normal net maternity function, part of which must lie above the negative age-axis. We are therefore solving an equation of the form

$$f_1(r) + f_2(r) = 1, \qquad \text{(I)}$$

where

$$f_1(r) = \frac{R_0}{\sigma\sqrt{(2\pi)}}\int_0^\infty \exp\left\{-\frac{1}{2}\left(\frac{x-\mu}{\sigma}\right)^2 - rx\right\} dx,$$

and

$$f_2(r) = \frac{R_0}{\sigma\sqrt{(2\pi)}}\int_0^\infty \exp\left\{-\frac{1}{2}\left(\frac{x+\mu}{\sigma}\right)^2 + rx\right\} dx.$$

It is not difficult to see that $f_1(r)$ is a monotonic decreasing function of r and that $f_2(r)$ is a monotonic increasing function of r. The function $f_1(r)$ is negligible for large r, and $f_2(r)$ is negligible except for very large r. We deduce that two real roots exist for equation (I) and these roots are approximately the roots of the two equations: $f_1(r) = 1$ and $f_2(r) = 1$. We require the solution of the equation $f_1(r) = 1$, and we therefore select the smaller root of equation (3.4.5).

2 All the derivatives of $z(r)$ with respect to r exist and are finite, and $z(r)$ is a monotonic strictly increasing function of r with a zero at $r = 0$. An inverse function $r(z)$ exists, and

$$\frac{dr}{dz} = \left(\frac{dz}{dr}\right)^{-1}.$$

The derivatives, d^2r/dz^2, d^3r/dz^3, d^4r/dz^4, and d^5r/dz^5 are found by successive differentiation, and the Taylor series takes the form

$$r(z) = r(0) + zr'(0) + \ldots + \frac{z^4}{4!}r^{iv}(0) + \frac{z^5}{5!}r^{v}(\theta z),$$

where $0 \leqslant \theta \leqslant 1$. The derivatives of r with respect to z in the right-hand side of this expression are the derivatives mentioned above evaluated at 0 and θz. The series is equivalent to formula (3.4.10) when z is set equal to $\log R_0$.

3 (i) For the stable population, $B(t) = A_0 e^{r_0 t}$. When $B(t+T)/B(t)$ is equated to R_0, we obtain the required formula.
(ii) The ratio $r_0/\log R_0$ is readily available from formula (3.4.10). The generation length T is found by taking the reciprocal.

4 $$\frac{\partial T}{\partial R_0} = -\frac{\kappa_2}{2R_0\kappa_1^2} + \text{(terms of smaller order)}.$$

$$\frac{\partial T}{\partial \kappa_n} = \frac{1}{n!}\left(-\frac{\log R_0}{\kappa_1}\right)^{n-1} + \text{(terms of smaller order)}.$$

5 Consider the integral $\displaystyle\int_0^\infty e^{-rx}\, e^{-kx}\left(\frac{x^5}{60}+\frac{x^3}{2}\right) dx.$

It has an infinite value for $r \leqslant -k$, and approaches zero as $r \to \infty$. It is a continuous function of r in the range $-k < r < \infty$ and is monotonically strictly decreasing. A unique real root r_0 therefore exists and it lies in the range $(-k, \infty)$. It is easy to prove that the integral is equal to

$$1 + \left\{\frac{2}{(r+k)^2} - 1\right\}\left\{\frac{1}{(r+k)^2} + 1\right\}^2,$$

so that the unique root satisfies the equation

$$2/(r+k)^2 = 1.$$

This equation has two real roots, and we select the one lying in the range $(-k, \infty)$, namely $r_0 = -k + \sqrt{2}$. *Note.* It would appear at first sight that repeated complex roots $-k \pm i$ exist for this population. No such roots exist, and the integral will not converge for such values of r.

6 From question 12 of section 2.14, ${}_xp_0 = e^{-\mu x}$. So we need to solve the equation

$$\int_0^\infty e^{-rx}\, e^{-\mu x}\, \lambda\, dx = 1.$$

That is, $\lambda/(\mu + r) = 1$. Only one root exists, the real root $r_0 = \lambda - \mu$.

7 In this case

$$B^*(r) = \frac{G^*(r)}{1-\phi^*(r)} = \frac{A_0}{r-r_0}+\frac{A_1}{r-r_1}+\frac{B_1}{(r-r_1)^2}+\frac{A_2}{r-r_2}+\ldots. \qquad \text{(II)}$$

The formulae of section (3.6) may be applied to evaluate $A_0, A_2, A_3, A_4, \ldots$. The only difficulties occur with A_1 and B_1. From equation (II) above,

$$(r-r_1)^2B^*(r) = \left(\frac{A_0}{r-r_0}+\frac{A_1}{r-r_1}+\frac{A_2}{r-r_2}+\ldots\right)(r-r_1)^2+B_1.$$

Therefore

$$B_1 = \lim_{r\to r_1} \frac{(r-r_1)^2G^*(r)}{1-\phi^*(r)}.$$

In a similar manner,

$$A_1 = \lim_{r\to r_1} \frac{\mathrm{d}}{\mathrm{d}r}\frac{(r-r_1)^2G^*(r)}{1-\phi^*(r)}.$$

The final formulae for A_1 and B_1 are rather messy.

8 The Laplace transform is

$$\frac{A_1}{r-r_1}+\frac{B_1}{(r-r_1)^2},$$

where Real $(r) >$ Real (r_1).
We now make use of equation (II) in the solution to question 7.

$$B(t) = \sum_{j=0}^{\infty} A_j\, e^{r_j t}+B_1 t\, e^{r_1 t}.$$

9 The probability density $(c^k x^{k-1} e^{-cx})/\Gamma(k)$ has mean k/c and variance k/c^2. We deduce that

$$c = (R_0R_1)/(R_0R_2-R_1^2) \quad \text{and} \quad k = R_1^2/(R_0R_2-R_1^2).$$

10 $\displaystyle\int_0^\infty e^{-rx}\,{}_xp_0\lambda(x)\,\mathrm{d}x = (R_0c^k)/(r+c)^k.$

Equating the integral to one, we obtain

$$r_n = c(R_0^{1/k}\, e^{(2n\pi i)/k}-1),$$

where $n = 0, \pm 1, \pm 2, \ldots$.
There is one difficulty however (similar to that encountered in exercise 5). The integral will only converge provided the real part of r is greater than $-c$. We therefore select those values of r_n for which the real part of $e^{(2n\pi i)/k}$ is positive.

11 That the Wicksell-population roots lie on a circle is obvious from the formula for r_n proved in question 10. The circle is centred on the real axis at the point $-c$ and its radius is $cR_0^{1/k}$.

12 $G(t) = N_t p_0\lambda(t)$.

$$A_j = N(c+r_j)/k.$$

13 Such a grand-daughter will be born if the newly-born female gives birth to a daughter at age $x(< y)$ and this daughter gives birth to a daughter at age $(y-x)$. The required probability is therefore

$$\int_{x=0}^{y} \frac{R_0 c^k x^{k-1} e^{-cx} dx}{\Gamma(k)} \frac{R_0 c^k (y-x)^{k-1} e^{-c(y-x)} dy}{\Gamma(k)}$$

which reduces to $(R_0^2 c^{2k} e^{-cy} y^{2k-1} dy)/\Gamma(2k)$.
The nth generation probability is $(R_0^n c^{nk} e^{-cy} y^{nk-1} dy)/\Gamma(nk)$.

14 The net maternity function is $(\lambda/\mu)\ (\mu^1 x^0 e^{-\mu x})/\Gamma(1)$.
A comparison of this formula with (3.7.6) indicates that the population is of the Wicksell type with $k = 1$ and $c = \mu$. The net reproduction rate R_0 is equal to λ/μ.

Chapter 4

1 The population will oscillate as follows:

Age-group	$t = 0$	$t = 1$	$t = 2$	$t = 3$
0–	1,000	6,000	2,000	1,000
1–	1,000	500	3,000	1,000
2–	1,000	$333\frac{1}{3}$	$166\frac{2}{3}$	1,000

The characteristic equation is $\lambda^3 - 1 = 0$, and the three latent roots are 1 and $\frac{1}{2}(-1 \pm i\sqrt{3})$. Clearly $|\lambda_0| = |\lambda_1| = |\lambda_2|$, and the population will oscillate indefinitely.

2 According to theorem 4.3.4, the second matrix is positive regular, but the first is not. These results may be confirmed by multiplication.

3 The characteristic equation is $\lambda^6 - 7\lambda^2 - 6 = 0$, with latent roots $\pm\sqrt{3}$, $\pm i\sqrt{2}$ and $\pm i$.

4 Both matrices have the same characteristic equation, namely $\lambda^3 - \lambda - 1 = 0$.

5 We need to compare the dominant latent root of the product matrix with the dominant latent roots of the squares of the matrices exhibited in question 4. The squared matrices both have the characteristic equation $\lambda^3 - 2\lambda^2 + \lambda - 1 = 0$, and the product matrix has as its characteristic equation $\lambda^3 - \frac{13}{6}\lambda^2 + \lambda - 1 = 0$. When the two polynomials are compared, we see that the population will increase at a faster rate in the oscillating environment.

6 The positive regular matrix

$$\begin{pmatrix} \frac{3}{4} & \frac{1}{4} \\ \frac{1}{2} & \frac{1}{2} \end{pmatrix}$$

has latent roots 1 and $\frac{1}{4}$.

7 The inverse of the Leslie matrix can be used for projecting a population backwards.

$$\begin{pmatrix} 0 & P_0^{-1} & 0 & 0 & \cdots & 0 \\ 0 & 0 & P_1^{-1} & 0 & \cdots & 0 \\ 0 & 0 & 0 & P_2^{-1} & \cdots & 0 \\ \cdot & \cdot & \cdot & \cdot & \cdots & \cdot \\ 0 & 0 & 0 & 0 & \cdots & P_{k-1}^{-1} \\ F_k^{-1} & -P_0^{-1}F_0F_k^{-1} & -P_1^{-1}F_1F_k^{-1} & -P_2^{-1}F_2F_k^{-1} & \cdots & -P_{k-1}^{-1}F_{k-1}F_k^{-1} \end{pmatrix}$$

8 It is soon apparent that the latent vectors of $\mathbf{A}^{-1}$ are the same as those of $\mathbf{A}$, but the latent roots are the reciprocals of those of $\mathbf{A}$. We conclude that $\mathbf{A}^{-1}$ has exactly one real positive root, and this root has the smallest modulus of all the latent roots of $\mathbf{A}^{-1}$. Thus a negative or complex root of $\mathbf{A}^{-1}$ is dominant, and furthermore this dominant root may be a repeated root.

9 We really need to examine the dominant latent roots of the two product matrices

$$\begin{pmatrix} 0 & 2 & 3 \\ \frac{1}{2} & 0 & 0 \\ 0 & \frac{2}{3} & 0 \end{pmatrix} \begin{pmatrix} \frac{1}{2} & 0 & 0 \\ 0 & \frac{1}{3} & 0 \\ 0 & 0 & \frac{1}{4} \end{pmatrix} \begin{pmatrix} 0 & 3 & 4 \\ \frac{1}{3} & 0 & 0 \\ 0 & \frac{3}{4} & 0 \end{pmatrix} = \begin{pmatrix} \frac{2}{9} & \frac{9}{16} & 0 \\ 0 & \frac{3}{4} & 1 \\ \frac{2}{27} & 0 & 0 \end{pmatrix},$$

and

$$\begin{pmatrix} 0 & 3 & 4 \\ \frac{1}{3} & 0 & 0 \\ 0 & \frac{3}{4} & 0 \end{pmatrix} \begin{pmatrix} \frac{1}{2} & 0 & 0 \\ 0 & \frac{1}{3} & 0 \\ 0 & 0 & \frac{1}{4} \end{pmatrix} \begin{pmatrix} 0 & 2 & 3 \\ \frac{1}{2} & 0 & 0 \\ 0 & \frac{2}{3} & 0 \end{pmatrix} = \begin{pmatrix} \frac{1}{2} & \frac{2}{3} & 0 \\ 0 & \frac{1}{3} & \frac{1}{2} \\ \frac{1}{8} & 0 & 0 \end{pmatrix}.$$

The respective characteristic equations are

$$\lambda^3 - \tfrac{35}{36}\lambda^2 + \tfrac{1}{6}\lambda - \tfrac{1}{24} = 0,$$

and

$$\lambda^3 - \tfrac{5}{6}\lambda^2 + \tfrac{1}{6}\lambda - \tfrac{1}{24} = 0.$$

A comparison of the two polynomials shows that the extermination procedure should be more effective if it is applied in the spring.
It should be noted that without the extermination procedure, the dominant latent roots are greater than one, and that with the procedure, both dominant latent roots are less than one.

10 The life table function p_x was defined in section 2.5, and values are tabulated in the Australian national life tables. The most convenient value for P_x is

$$P_x = \tfrac{1}{2}(p_x + p_{x+1}).$$

11 (i) $(\Delta\lambda_0)/\lambda_0 \doteqdot (0.01\ \lambda_0^{\kappa_1})/(R_0\kappa_1) \doteqdot 0.01\ \kappa_1^{-1}$.

$$\Delta\lambda_0 \doteqdot 0.01\ \kappa_1^{-1}, \text{ since } \lambda_0 \text{ is close to one.}$$

(ii) $\Delta R_0 = 0.01\ R_0$.
(iii) We note that $r_0 \doteqdot \lambda_0 - 1$, so that $\partial r_0/\partial R_0 \doteqdot \partial\lambda_0/\partial R_0$. Using equation (3.5.1) therefore,

$$\Delta\lambda_0 \doteqdot (\Delta R_0)/(R_0\kappa_1) = 0.01\ \kappa_1^{-1}.$$

12 The latent roots of the matrix are both equal to 0.5. There exists a matrix $\mathbf{H}$ such that

$$\mathbf{H}^{-1}\mathbf{A}\mathbf{H} = \begin{pmatrix} 0.5 & 1 \\ 0 & 0.5 \end{pmatrix} \qquad \text{(section 4.5).}$$

If we make the transformation $\mathbf{n}(t) = \mathbf{H}^{-1}\mathbf{N}(t)$, we find that

$$n_2(t) = \alpha_2 \, e^{0.5t} \quad \text{and} \quad n_1(t) = (\alpha_1 + \alpha_2 t)\, e^{0.5t}.$$

Finally,

$$N_1(t) = (100 - 5t)\, e^{0.5t};$$

and

$$N_2(t) = (100 - 10t)\, e^{0.5t}.$$

These results are valid when $0 \leqslant t \leqslant 10$.

13 Yes. (Use theorem 4.3.4).

Chapter 5

2 If we denote the population size at time t by $N(t)$,

$$\mathscr{E}N(t) = e^{\lambda t} \quad \text{and} \quad \operatorname{Var} N(t) = e^{\lambda t}(e^{\lambda t} - 1),$$

assuming an initial population of one.

5 When $\lambda = \mu$, equation (5.4.6) becomes

$$\frac{d}{dt} M_2(t) = (\lambda + \mu),$$

and we deduce that

$$M_2(t) = 1 + (\lambda + \mu)\, t.$$

It follows that

$$\operatorname{Var} N(t) = (\lambda + \mu)\, t, \; (\lambda = \mu).$$

6 Use the method of section 5.4. We find that

$$dt = \frac{-dz}{\lambda(z-1)^2}$$

so that

$$t - \frac{1}{\lambda(z-1)} = C,$$

and C is a function of the height of the contour. Finally

$$\phi(z, t) = \frac{(\lambda t - 1)(z-1) - 1}{\lambda t(z-1) - 1}.$$

7 Differentiate $\phi(z, t)$ partially with respect to z twice. The mean is equal to the first partial derivative when z is set equal to one, and the second (falling) factorial moment is equal to the second derivative when z is set equal to one. We find that

$$\mathscr{E}N(t) = 1 \quad \text{and} \quad \mathscr{E}N(t)\{N(t) - 1\} = 2\lambda t.$$

Therefore $\operatorname{Var} N(t) = 2\lambda t = (\lambda + \mu)\, t$, since $\lambda = \mu$.

8 The probability-generating function $\phi(z, t)$ is known from question 6. The probability that the population is extinct at time t is $\phi(0, t)$ which is equal to $\{1+(\lambda t)^{-1}\}^{-1}$. The probability of ultimate extinction is found by considering the limit as t tends to ∞, and the limit is one.

9 Rather than differentiate (5.5.20) partially with respect to z twice, it is possibly easier to multiply equation (5.5.2) by n^2 and sum for all non-negative values of n. We find that

$$\frac{d}{dt} M_2(t) = 2(\lambda-\mu)\, M_2(t)+(\lambda+\mu+2\kappa)\, M_1(t)+\kappa.$$

The variance turns out to be

$$\kappa\{\lambda\, e^{(\lambda-\mu)t}-\mu\}\,\{e^{(\lambda-\mu)t}-1\}/(\lambda-\mu)^2.$$

10 Use the method of section 5.5. Equations (5.5.16) become

$$\frac{dz}{\lambda(z-1)^2} = \frac{dt}{-1} = \frac{d\phi}{-\kappa(z-1)\,\phi},$$

with solutions
$$\begin{cases} t-\dfrac{1}{\lambda(z-1)} = C; \\ (z-1)^{\kappa}\phi^{\lambda} = K. \end{cases}$$

At time $t = 0$, the population is assumed to have zero members, so that $\phi(z, 0) = 1$. Finally

$$\phi(z, t) = \{1-\lambda(z-1)\,t\}^{-\kappa/\lambda}.$$

11 Use the partial derivatives with respect to z of $\phi(z, t)$ in question 10. The mean is κt and the variance $\kappa t(1+\frac{1}{2}\lambda t+\frac{1}{2}\mu t)$.

12 There are always immigrants entering the population, and so the probability of ultimate extinction is zero. Note however, that the population may have zero members at any time in the future.

14 The probability-generating function (5.6.2) may be written in the form

$$\left(1-\frac{\lambda}{\mu}\right)^{\kappa/\lambda}\left[1+\frac{\kappa}{\lambda}\left\{\frac{\lambda}{\mu}z+\left(\frac{\lambda}{\mu}\right)^2\frac{z^2}{2}+\left(\frac{\lambda}{\mu}\right)^3\frac{z^3}{3}+\dots\right\}\right]$$

provided κ/λ is very small.

The conditional distribution of the population size $N(t)$ given that it is non-zero is found by omitting the term in z^0 in the above formula and multiplying all the other terms by a constant in order that the conditional probabilities sum to one. We obtain the formula quoted in the question.

15 A formula for $\phi(z, t)$ is given by equation (5.5.20). The probability that the population has zero members at time t is found by setting $\lambda = 1$, $\mu = 2$, $\kappa = 1$ and $z = 0$ in equation (5.5.20). It is equal to $(2-e^{-t})^{-1}$ which tends to one half as $t \to \infty$.

16 We set $\lambda = 1$, $\mu = 2$, $\kappa = 1$ and $z = 0$ in formula (5.6.2). The probability is one half.

17 Use the method of section 5.3 to obtain

$$P_n(t) = \binom{n-1}{m-1} e^{-m\lambda t} (1 - e^{-\lambda t})^{n-m}$$

where $n = m, m+1, m+2, \ldots$

18 The results quoted in section 5.4 and in the solution to question 8 are applicable to this type of population when there is a single initial ancestor. With this model, co-existing individuals are assumed to act independently. We deduce therefore that the probability of extinction for a population with N_0 initial ancestors is 1 when $\mu \geqslant \lambda$ and $(\mu/\lambda)^{N_0}$ if $\lambda > \mu$.

Chapter 6

4 The latent roots are $\pm\sqrt{\tfrac{3}{2}}$.

5 $K_{t+1}(\theta_0, \theta_1) = K_t(\log(\tfrac{1}{2}+\tfrac{1}{2}e^{\theta_1}), 3\theta_0)$. (6.2.4)

$$\left(\begin{array}{c}\mathbf{e}_{t+1}\\ \hline \mathbf{C}_{t+1}\end{array}\right) = \left(\begin{array}{cc|cccc} 0 & 3 & 0 & 0 & 0 & 0\\ \tfrac{1}{2} & 0 & 0 & 0 & 0 & 0\\ \hline 0 & 0 & 0 & 0 & 0 & 9\\ 0 & 0 & 0 & 0 & \tfrac{3}{2} & 0\\ 0 & 0 & 0 & \tfrac{3}{2} & 0 & 0\\ \tfrac{1}{4} & 0 & \tfrac{1}{4} & 0 & 0 & 0 \end{array}\right) \left(\begin{array}{c}\mathbf{e}_t\\ \hline \mathbf{C}_t\end{array}\right) \qquad (6.2.5)$$

$\phi_{t+1}(z_0, z_1) = \phi_t((\tfrac{1}{2}+\tfrac{1}{2}z_1), z_0^3)$ (6.2.6)

The probability of ultimate extinction is a root of the cubic equation $x = \tfrac{1}{2}+\tfrac{1}{2}x^3$. The appropriate root is $x = \tfrac{1}{2}(\sqrt{5}-1)$.

6 $K_{t+1}(\theta_0, \theta_1) = K_t(\log(\tfrac{1}{2}+\tfrac{1}{2}e^{\theta_1}), 6\log(\tfrac{1}{2}+\tfrac{1}{2}e^{\theta_0}))$. (6.2.4)

$$\left(\begin{array}{c}\mathbf{e}_{t+1}\\ \hline \mathbf{C}_{t+1}\end{array}\right) = \left(\begin{array}{cc|cccc} 0 & 3 & 0 & 0 & 0 & 0\\ \tfrac{1}{2} & 0 & 0 & 0 & 0 & 0\\ \hline 0 & \tfrac{3}{2} & 0 & 0 & 0 & 9\\ 0 & 0 & 0 & 0 & \tfrac{3}{2} & 0\\ 0 & 0 & 0 & \tfrac{3}{2} & 0 & 0\\ \tfrac{1}{4} & 0 & \tfrac{1}{4} & 0 & 0 & 0 \end{array}\right) \left(\begin{array}{c}\mathbf{e}_t\\ \hline \mathbf{C}_t\end{array}\right) \qquad (6.2.5)$$

$\phi_{t+1}(z_0, z_1) = \phi_t((\tfrac{1}{2}+\tfrac{1}{2}z_1), (\tfrac{1}{2}+\tfrac{1}{2}z_0)^6)$. (6.2.6)

The probability of ultimate extinction is given by the root of the equation $x = \tfrac{1}{2}+\tfrac{1}{2}x^6$ lying in the range $0 < x < 1$. We find that the appropriate root is $x = 0.50866$.

7 According to the theory of section 5.5, we must solve the auxiliary equations

$$\frac{dx}{1} = \frac{dt}{1} = -\frac{d\alpha}{\mu\alpha}$$

and these equations have solutions

$$t - x = c \quad \text{and} \quad \alpha = K\exp\left(-\int_0^x \mu_x dx\right).$$

We conclude that the general solution is of the form

$$\alpha(x, t) = \exp\left(-\int_0^x \mu_x \mathrm{d}x\right) \phi(t-x).$$

This is consistent with the Lotka equation (3.2.1).

8 Select the coefficients of $\theta(0)$ and $\{\theta(0)\}^2$ from the expansion of formula (6.4.4). We find that

$$\alpha(0, t) = \int_0^\infty \lambda(x)\, \alpha(x, t)\, \mathrm{d}x$$

and $\beta(0, t)$ is also equal to the same integral. The above equation for $\alpha(0, t)$ is essentially equation (3.2.2).

9 $\mathscr{E}M'(M'-1)(M'-2) = p^3\mathscr{E}M(M-1)(M-2)$.

Chapter 7

1 The proof is analogous to that given in the final paragraph of section 7.8.

2 $r_0(\text{males}) = X\lambda_{mf}-\mu_m$; $r_0(\text{females}) = X^{-1}\lambda_{fm}-\mu_f$.

$s_0 = \frac{1}{2}[-(\mu_m+\mu_f)+\{(\mu_m-\mu_f)^2+4\lambda_{mf}\lambda_{fm}\}^{\frac{1}{2}}]$.

We select the positive square root because the double integral will only converge when s_0 is greater than the maximum of $-\mu_m$ and $-\mu_f$.

4 $s_0 = c\{(M_0 F_0)^{1/(k+j)}-1\}$.

5 Define $m_0 = r_0(\text{males})$; $f_0 = r_0(\text{females})$;

$A = \kappa_1(\text{males})$; $a = \kappa_1(\text{females})$;

$B = \kappa_2(\text{males})$; $b = \kappa_2(\text{females})$.

Then $(A+a)\, s_0+\frac{1}{2}(B+b)\, s_0^2 = Am_0+af_0+\frac{1}{2}Bm_0^2+\frac{1}{2}bf_0^2$, a quadratic equation for s_0 in terms of m_0 and f_0. We select the positive square root and expand the surd by the binomial theorem.

6 The following integral equation is obtained:

$$X^{K-L}\int_\alpha^\beta \cdots \int_\alpha^\beta \exp\left(-s\sum_{i=1}^{K+L} x_i\right)\left\{\prod_{i=1}^{K} \phi(x_i)\right\}\left\{\prod_{K+1}^{K+L} \xi(x_i)\right\} \mathrm{d}x_1 \ldots \mathrm{d}x_{K+L} = 1.$$

There is exactly one real solution, and the complex roots occur in conjugate pairs.

8 $s_0 \doteqdot \dfrac{K\kappa_1(\text{males})\, r_0(\text{males})+L\kappa_1(\text{females})\, r_0(\text{females})}{K\kappa_1(\text{males})+L\kappa_1(\text{females})}$.

9 Arithmetic: -0.0528; Geometric: -0.0529; Harmonic: -0.0530.

10

	Stable population		
	M	*F*	*C*
Arithmetic	1,000	1,205	1,042
Geometric	1,000	1,206	1,039
Harmonic	1,000	1,206	1,035

11 $\mu t = -\log\{1-\alpha/M(0)\}$.

Chapter 8

1 For Lotka's population, $p_0 = 0.4982$. We deduce that

$$\frac{b}{1-c} = \sum_{j=1}^{\infty} p_j = 0.5018.$$

It is easy to prove that the mean $m = b/(1-c)^2 = 1.1450$.
From these two equations we deduce that $b = 0.2199$, and $c = 0.5617$.
The following values of p_j are obtained:

$$p_0 = 0.4982;\quad p_1 = 0.2199;\quad p_2 = 0.1235;\quad p_3 = 0.0694;$$

$$p_4 = 0.0390;\quad p_5 = 0.0219;\quad p_6 = 0.0123;\quad p_7 = 0.0069;$$

$$p_8 = 0.0039;\quad p_9 = 0.0022;\quad p_{10} = 0.0012;\quad p_{11} = 0.0007.$$

(The fit might be improved by adjusting p_0.) According to the theory of section 8.5, the probability of ultimate extinction is given by

$$q = z_0 = \frac{1-b-c}{c(1-c)} = 0.8871.$$

2 The mean m is one, and the probability of ultimate extinction is therefore one. Using formula (8.5.3), we see that $b = (1-c)^2$, and therefore from equation (8.5.2), we know that

$$f(z) = \{c-(2c-1)\,z\}/(1-cz) \quad (m = 1).$$

It is easy to prove by induction that

$$f_n(z) = \{nc-(nc+c-1)\,z\}/(1-c+nc-ncz) \quad (m = 1).$$

3 $f^0(z) = \frac{1}{2}+\frac{1}{2}z_1$; $f^1(z) = z_0^3$. $\quad \mathbf{M} = \begin{pmatrix} 0 & 3 \\ \frac{1}{2} & 0 \end{pmatrix}$.

The latent roots of $\mathbf{M}$ are known to be $\pm\sqrt{\frac{3}{2}}$, and the probabilities of ultimate extinction are therefore strictly less than one. We need to solve the simultaneous equations

$$z_0 = \tfrac{1}{2}+\tfrac{1}{2}z_1 \quad \text{and} \quad z_1 = z_0^3.$$

Eliminating z_1, we must solve the cubic equation $z_0 = \frac{1}{2}+\frac{1}{2}z_0^3$, and we find that $z_0 = \frac{1}{2}(\sqrt{5}-1) = 0.618034$.

4 $\frac{1}{8}(\sqrt{5}-1)^3$ or 0.23606.

5 $f_2^1(z_1, z_2) = \dfrac{(ad+ab+cg)+(ae+b^2+ch)\,z_1+(af+bc+ci)\,z_2}{(d^2+ae+fg)+(de+be+fh)\,z_1+(df+ec+fi)\,z_2}.$

A similar expression may be written down for $f_2^2(z_1, z_2)$, and the two denominators are again identical. The general case may be proved by induction.

6

Generation	1	2	3	4	5	6
Probability	0.4982	0.6509	0.7240	0.7667	0.7945	0.8138
Generation	7	8	9	10	∞	
Probability	0.8279	0.8387	0.8470	0.8536	0.8871	

7 $\log f(z) = \log f(1) + \log z \left[\dfrac{d \log f}{d \log z}\right]_{z=1} + \frac{1}{2}(\log z)^2 \left[\dfrac{d^2 \log f}{d(\log z)^2}\right]_{z=1} + \ldots.$

It is a straightforward matter to prove that

$$\left[\frac{d \log f}{d \log z}\right]_{z=1} = m \quad \text{and} \quad \left[\frac{d^2 \log f}{d(\log z)^2}\right]_{z=1} = \sigma^2.$$

But z must be equal to $f(z)$, so that

$$\log z \doteqdot m \log z + \tfrac{1}{2}\sigma^2(\log z)^2$$

and Bartlett's formula follows.

8 Set σ^2 equal to m. The extinction probability is

$$e^{-2(m-1)/m} = e^{-2\delta/(1+\delta)} \doteqdot 1 - 2\delta + 4\delta^2.$$

Chapter 9

2 We shall make use of the notation of chapters 4 and 9. The probability that a female now in age group 0 will die aged x last birthday is $P_0 P_1 \ldots P_{x-1} Q_x$. Given that she will die aged x last birthday, the probability that she will have exactly j daughters is the coefficient of z^j in

$$\prod_{r=0}^{x} (G_r + F_r z).$$

The probability that a female now in age group 0 will have exactly j daughters is therefore the coefficient of z^j in

$$f(z) = \sum_{x=0}^{k} Q_x(G_x + F_x z)\left\{\prod_{r=0}^{x-1} P_r(G_r + F_r z)\right\}.$$

If we now apply a generation approach to this female, we know the probability that she will have exactly j children, and this probability law will also apply to her daughters. We can therefore determine the probability of ultimate extinction of the population by considering a one-type Galton–Watson process and solving the equation $f(z) = z$.

3 Let us expand the general formula given in the solution to question 2 for the two-type case. We obtain

$$f(z) = (Q_0 G_0 + P_0 Q_1 G_0 G_1) + (Q_0 F_0 + P_0 Q_1 G_1 F_0 + P_0 Q_1 G_0 F_1)\,z + (P_0 Q_1 F_0 F_1)\,z^2.$$

For the numerical example, $f(z) = 0.1625 + 0.5000z + 0.3375z^2$, and we find that the probability of ultimate extinction is $\frac{13}{27}$ or 0.48148. Let us now find the probability of ultimate extinction using the two-type process. We need to solve the equations

$$z_0 = f^0(z_0, z_1) = Q_0G_0 + Q_0F_0z_0 + P_0G_0z_1 + P_0F_0z_0z_1,$$

and

$$z_1 = f^1(z_0, z_1) = Q_1G_1 + Q_1F_1z_0.$$

The second equation gives us z_1 in terms of z_0 and this may be substituted into the first equation. We obtain a quadratic equation for z_0 which is identical with the one-type quadratic.

4 The solution of the quadratic equation in the solution to question 3 is $z_0 = \frac{13}{27}$. We substitute this in the second of the two basic equations to obtain $z_1 = \frac{11}{18}$. The probability of ultimate extinction is therefore

$$z_0^5 z_1^3 = (\tfrac{13}{27})^5 (\tfrac{11}{18})^3 = 0.00591.$$

5 $$\mathbf{P} = \begin{pmatrix} 0.05 & 0 \\ 0.45 & 0 \\ 0.45 & 0 \\ 0 & 0.75 \end{pmatrix}; \quad \begin{array}{l} \mathbf{Q} = \begin{pmatrix} 1 & 1 & 0 & 1 \\ 0 & 1 & 1 & 0 \end{pmatrix}, \\ \mathbf{QP} = \begin{pmatrix} 0.5 & 0.75 \\ 0.9 & 0 \end{pmatrix} \end{array}; \quad \mathbf{D}_{21} = \begin{pmatrix} 0.25 & 0.1875 \\ 0 & 0 \\ 0 & 0 \\ 0.09 & 0 \end{pmatrix}.$$

The matrix $\mathbf{D}_{21}$ should be checked against the submatrix $\mathbf{D}$ in equation (9.3.11).

Chapter 10

1 $$\mathbf{N} = \left(\begin{array}{c|c} a_1 & 0 \\ a_2 & 0 \\ \hline a_1(1-a_1) & a_1a_1 \\ -a_1a_2 & a_1a_2 \\ -a_2a_1 & a_2a_1 \\ a_2(1-a_2) & a_2a_2 \end{array}\right).$$

2 The dominant latent root is 1. (The other latent root is $-\frac{1}{2}$.) When the population develops a stable age distribution, four sevenths of the population will be in age group zero.

4 Let us denote the moment vector at time t by $\mathbf{m}(t)$, and recall the results of question 3. It is possible to prove that the moment vectors obey a linear recurrence relation of the form

$$\mathbf{m}(t+1) = \mathbf{T}_2\mathbf{M}_2\mathbf{B}_2\mathbf{F}_2^*\mathbf{T}_1\mathbf{M}_1\mathbf{B}_1\mathbf{F}_1\mathbf{T}_0\mathbf{m}(t).$$

The matrix $\mathbf{T}_0$ corresponds to a linear transformation which yields three random variables: those aged zero, those aged one, and the total population, instead of the two original variables. The matrix $\mathbf{F}_1$ gives us the falling factorial moments of these three random variables. The matrix $\mathbf{B}_1$ corresponds to a branching process in which the individuals aged 0 and 1 survive and reproduce according to the probabilities in question 2 and the total population variable survives with probability one. Premultiplication by the matrices $\mathbf{T}_1$ and $\mathbf{M}_1$ gives us the moments of three random variables: the numbers of individuals aged 0 and 1 at time $t+1$ without the effects of immigration and the total population at time t. We now deal with immigration. The first two random variables must have their moments converted into falling factorial moments to account for their survival with probability one; the total population random variable does not need this adjustment according to question 3. It follows that $\mathbf{F}_2^*$ is slightly different from the usual $\mathbf{F}$ matrix: all the elements of the row in the lower left-hand submatrix corresponding to the second-order moment of the total population at time t are zero. The matrices $\mathbf{T}_2$, $\mathbf{M}_2$ and $\mathbf{B}_2$ then account for immigration and the survivors and descendants of those in the population at time t and the recurrence matrix for first- and second-order moments is found to be

$$\left(\begin{array}{cc|cccc} \frac{5}{6} & 1 & 0 & 0 & 0 & 0 \\ \frac{11}{12} & \frac{1}{6} & 0 & 0 & 0 & 0 \\ \hline \frac{7}{12} & \frac{5}{9} & \frac{25}{36} & \frac{5}{6} & \frac{5}{6} & 1 \\ 0 & 0 & \frac{55}{72} & \frac{5}{36} & \frac{11}{12} & \frac{1}{6} \\ 0 & 0 & \frac{55}{72} & \frac{11}{12} & \frac{5}{36} & \frac{1}{6} \\ \frac{17}{48} & \frac{1}{6} & \frac{121}{144} & \frac{11}{72} & \frac{11}{72} & \frac{1}{36} \end{array}\right).$$

5 The dominant latent root is $\frac{1}{6}(3+\sqrt{37})$. This is greater than $1\frac{1}{2}$, which is obtained by adding one half to the dominant latent root of the population in question 2. If the proportion of the immigrants in age group 0 is altered, the dominant latent root will change also; it may lie above or below $1\frac{1}{2}$.

6 The expectation matrix is

$$\begin{pmatrix} \frac{11}{14} & \frac{20}{21} \\ \frac{27}{28} & \frac{3}{14} \end{pmatrix} = \begin{pmatrix} \frac{1}{2} & \frac{2}{3} \\ \frac{3}{4} & 0 \end{pmatrix} + \frac{1}{2} \begin{pmatrix} \frac{4}{7} & \frac{4}{7} \\ \frac{3}{7} & \frac{3}{7} \end{pmatrix}.$$

The dominant latent root of this matrix is 1.5 which agrees with our intuition! It is easy to see that the corresponding latent vector is proportional to the stable age distribution of question 2.

The Australian life table (males) 1961

Age x	l_x	d_x	p_x	q_x	μ_x	$\mathring{e}_x$
0	100,000	2,239	0.97761	0.02239	—	67.92
1	97,761	177	0.99819	0.00181	—	68.46
2	97,584	117	0.99880	0.00120	—	67.59
3	97,467	88	0.99910	0.00090	0.00102	66.67
4	97,379	64	0.99934	0.00066	0.00076	65.73
5	97,315	56	0.99942	0.00058	0.00060	64.77
6	97,259	53	0.99945	0.00055	0.00056	63.81
7	97,206	52	0.99947	0.00053	0.00054	62.84
8	97,154	49	0.99950	0.00050	0.00052	61.87
9	97,105	43	0.99956	0.00044	0.00047	60.91
10	97,062	40	0.99959	0.00041	0.00042	59.93
11	97,022	41	0.99958	0.00042	0.00041	58.96
12	96,981	45	0.99954	0.00046	0.00044	57.98
13	96,936	51	0.99947	0.00053	0.00049	57.01
14	96,885	60	0.99938	0.00062	0.00057	56.04
15	96,825	73	0.99925	0.00075	0.00068	55.07
16	96,752	92	0.99905	0.00095	0.00084	54.11
17	96,660	119	0.99877	0.00123	0.00107	53.16
18	96,541	157	0.99837	0.00163	0.00144	52.23
19	96,384	169	0.99825	0.00175	0.00173	51.31
20	96,215	166	0.99827	0.00173	0.00175	50.40
21	96,049	163	0.99830	0.00170	0.00172	49.49
22	95,886	158	0.99835	0.00165	0.00168	48.57
23	95,728	151	0.99842	0.00158	0.00162	47.65
24	95,577	145	0.99848	0.00152	0.00155	46.73
25	95,432	140	0.99853	0.00147	0.00149	45.80
26	95,292	138	0.99855	0.00145	0.00146	44.86
27	95,154	140	0.99853	0.00147	0.00146	43.93
28	95,014	143	0.99850	0.00150	0.00149	42.99
29	94,871	145	0.99847	0.00153	0.00152	42.06
30	94,726	149	0.99843	0.00157	0.00155	41.12
31	94,577	152	0.99839	0.00161	0.00159	40.18
32	94,425	158	0.99833	0.00167	0.00164	39.25
33	94,267	164	0.99826	0.00174	0.00170	38.31
34	94,103	172	0.99817	0.00183	0.00178	37.38

Age x	l_x	d_x	p_x	q_x	μ_x	$\mathring{e}_x$
35	93,931	182	0.99806	0.00194	0.00188	36.45
36	93,749	195	0.99792	0.00208	0.00201	35.51
37	93,554	211	0.99774	0.00226	0.00217	34.59
38	93,343	231	0.99753	0.00247	0.00236	33.67
39	93,112	253	0.99728	0.00272	0.00259	32.75
40	92,859	279	0.99700	0.00300	0.00286	31.84
41	92,580	306	0.99670	0.00330	0.00315	30.93
42	92,274	336	0.99636	0.00364	0.00347	30.03
43	91,938	369	0.99599	0.00401	0.00383	29.14
44	91,569	404	0.99559	0.00441	0.00421	28.25
45	91,165	442	0.99515	0.00485	0.00463	27.38
46	90,723	485	0.99465	0.00535	0.00510	26.51
47	90,238	533	0.99409	0.00591	0.00564	25.65
48	89,705	587	0.99346	0.00654	0.00623	24.80
49	89,118	645	0.99276	0.00724	0.00690	23.96
50	88,473	711	0.99196	0.00804	0.00765	23.13
51	87,762	783	0.99108	0.00892	0.00850	22.31
52	86,979	860	0.99011	0.00989	0.00943	21.51
53	86,119	944	0.98904	0.01096	0.01046	20.72
54	85,175	1,033	0.98787	0.01213	0.01160	19.94
55	84,142	1,127	0.98661	0.01339	0.01283	19.18
56	83,015	1,225	0.98524	0.01476	0.01415	18.43
57	81,790	1,331	0.98373	0.01627	0.01561	17.70
58	80,459	1,442	0.98208	0.01792	0.01722	16.99
59	79,017	1,561	0.98025	0.01975	0.01898	16.29
60	77,456	1,685	0.97824	0.02176	0.02094	15.60
61	75,771	1,817	0.97602	0.02398	0.02310	14.94
62	73,954	1,952	0.97360	0.02640	0.02548	14.29
63	72,002	2,087	0.97101	0.02899	0.02806	13.67
64	69,915	2,216	0.96830	0.03170	0.03079	13.06
65	67,699	2,338	0.96546	0.03454	0.03366	12.47
66	65,361	2,451	0.96250	0.03750	0.03666	11.90
67	62,910	2,557	0.95936	0.04064	0.03982	11.34
68	60,353	2,657	0.95598	0.04402	0.04320	10.80
69	57,696	2,752	0.95230	0.04770	0.04688	10.28
70	54,944	2,844	0.94823	0.05177	0.05094	9.77
71	52,100	2,932	0.94372	0.05628	0.05546	9.27
72	49,168	3,008	0.93882	0.06118	0.06046	8.80
73	46,160	3,068	0.93354	0.06646	0.06588	8.34
74	43,092	3,108	0.92788	0.07212	0.07174	7.90
75	39,984	3,124	0.92186	0.07814	0.07804	7.47
76	36,860	3,115	0.91549	0.08451	0.08474	7.06
77	33,745	3,084	0.90860	0.09140	0.09196	6.67
78	30,661	3,032	0.90112	0.09888	0.09985	6.29
79	27,629	2,960	0.89287	0.10713	0.10856	5.92

Age x	l_x	d_x	p_x	q_x	μ_x	$\mathring{e}_x$
80	24,669	2,866	0.88383	0.11617	0.11823	5.57
81	21,803	2,749	0.87393	0.12607	0.12894	5.24
82	19,054	2,606	0.86321	0.13679	0.14074	4.92
83	16,448	2,440	0.85164	0.14836	0.15366	4.63
84	14,008	2,250	0.83938	0.16062	0.16766	4.35
85	11,758	2,042	0.82637	0.17363	0.18272	4.08
86	9,716	1,819	0.81274	0.18726	0.19886	3.84
87	7,897	1,591	0.79849	0.20151	0.21602	3.61
88	6,306	1,363	0.78380	0.21620	0.23417	3.40
89	4,943	1,143	0.76867	0.23133	0.25321	3.20
90	3,800	938	0.75325	0.24675	0.27312	3.02
91	2,862	751	0.73759	0.26241	0.29377	2.85
92	2,111	587	0.72179	0.27821	0.31512	2.70
93	1,524	448	0.70585	0.29415	0.33711	2.55
94	1,076	334	0.68977	0.31023	0.35979	2.42
95	742	242	0.67351	0.32649	0.38322	2.29
96	500	171	0.65706	0.34294	0.40748	2.17
97	329	118	0.64037	0.35963	0.43267	2.06
98	211	79	0.62346	0.37654	0.45887	1.96
99	132	52	0.60636	0.39364	0.48610	1.86
100	80	33	0.58913	0.41087	—	—
101	47	20	0.57178	0.42822	—	—
102	27	12	0.55435	0.44565	—	—
103	15	7	0.53690	0.46310	—	—
104	8	4	0.51943	0.48057	—	—
105	4	2	0.50201	0.49799	—	—
106	2	1	0.48467	0.51533	—	—
107	1	1	0.46747	0.53253	—	—
108	—	—	0.45047	0.54953	—	—
109	—	—	0.43370	0.56630	—	—

Author Index

Subject Index

Milton Keynes UK
Ingram Content Group UK Ltd.
UKHW041522181024
449640UK00009B/150

9 780521 294423